6/18

BEST BEFORE

Also available in the Bloomsbury Sigma series:

BEST BEFORE

THE EVOLUTION AND FUTURE
OF PROCESSED FOOD

Nicola Temple

Bloomsbury Sigma
An imprint of Bloomsbury Publishing Plc

50 Bedford Square
London
WC1B 3DP
UK

1385 Broadway
New York
NY 10018
USA

www.bloomsbury.com

BLOOMSBURY and the Diana logo are trademarks of Bloomsbury Publishing Plc

First published 2018

British Library Cataloguing-in-Publication Data
A catalogue record for this book is available from the British Library.

Library of Congress Cataloguing-in-Publication data has been applied for.

ISBN (hardback) 978-1-4729-4143-5
ISBN (trade paperback) 978-1-4729-4144-2
ISBN (ebook) 978-1-4729-4140-4

2 4 6 8 10 9 7 5 3 1

Illustrations by Nicola Temple

Bloomsbury Sigma, Book Thirty-two

Typeset by Deanta Global Publishing Services, Chennai, India
Printed and bound in Great Britain by CPI Group (UK) Ltd, Croydon CR0 4YY

To find out more about our authors and books visit www.bloomsbury.com.
Here you will find extracts, author interviews, details of forthcoming events
and the option to sign up for our newsletters.

Contents

I'll Have That with a Side of Pragmatism Please

I grew up on a homestead in rural Ontario, Canada. Ten blissful acres on which to lose myself as a child. We had a few sheep, a couple of dozen chickens, a pig named Bessie, a dairy cow called Penny and so many raised garden beds that people driving by no doubt wondered whether our farm doubled as a graveyard. I even had a mallard duck named Slipper who had imprinted on me as a hatchling and followed me everywhere. We grew as much of our own food as we could. And it was here that I first learned about processing food.

For several weekends throughout the summer and autumn, our small kitchen would be converted into a food-processing facility, filled with produce from the garden. No surface was left uncovered. There were pots of water boiling on the stove top and bowls of ice-cold water lined up along the wooden butcher-block counter from the stove top all the way to our Harvest Gold fridge. Freezer bags were at the ready and colanders, knives and chopping boards were in abundance. My mum and her friend Maureen, my grandmother and I, would be poised to process the garden's bounty so that we would have fruit and vegetables throughout the winter and spring.

My baby sister was plunked into one of those gravity-defying high-chairs that attached right to the table with suction cups and we set to work like a well-oiled machine. My mum worked the stove and the ice bowls, while Maureen and my grandmother prepared the veg. I was relegated to writing the name and date on the freezer bag (as if anyone couldn't tell it was corn in that clear bag) until I was responsible enough to wield a knife. My sister's job was to consume any vegetable product that fell within her grasp.

We blanched cob after cob of corn, trimming it off the cob (my grandma always cut too deep into the cob) and packaging it up for freezing. Cucumbers and onions were pickled. Beans were trimmed and peas were shucked and all were blanched and frozen. Fruit was frozen, pureed or preserved as jams. By midday, the floor would be sticky with berries and corn starch, there would be corn silk *everywhere* and we would all be dripping in sweat from the humidity of the boiling pots. If we stood to work, our backs ached. If we sat, our thighs and sweat-soaked backs would stick to the orange vinyl kitchen chairs (oh yes, we were terribly in vogue). And yet, this never felt like work. This was a social event. Particularly if a bottle of wine was opened in the afternoon and especially if my grandmother had left early. It was then that my mum and her friend would bring up outrageous stories from the past while I tried to make myself invisible in the corner, chopping in silence. It was ... enlightening. And, like most of my great memories in life, food was the centrepiece.

I had a privileged childhood. We didn't have a lot of money, but we had plenty of good food. I knew, sometimes in graphic detail, where good food came from and how it was prepared. My sister, who was either less pragmatic or more empathetic than me (probably both), became a lifelong vegetarian under the weight of this intimate knowledge.

As I sought autonomy, I rebelled against anything wholemeal. In school I envied the kids who had sandwiches made with thin, tasteless, white, fluffy pre-sliced bread. Mine was brown, dense and could easily have been used to hold the classroom door open. It was a revelation when I learned that pizza didn't have to have a two-inch-thick rectangular wholemeal crust!

When I had an income of my own, I took every opportunity to sample ready meals and fast food. It was bliss. I lived above a fish and chip shop for a while for goodness' sake, I didn't stand a chance! Luckily, any subtle nuances to my health as a result of my poor diet were clearly masked by the overwhelming flavour of youth.

I returned to my wholesome food roots eventually.

When I started to do extensive research for my first book, entitled *Sorting the Beef from the Bull: The Science of Food Fraud Forensics*, which I co-authored with Professor Richard Evershed, I learned of food horror stories I never thought possible. I read about plastic rice, fake eggs and putrid meat dipped in bleach and re-directed into the human food chain. I learned about fake milk made with shampoo and urea, and sawdust coloured with toxic dyes to resemble ground spices. Horse meat in beef lasagne was tame by comparison. These were all clearly illegal and shocking examples of how criminals adulterate food for the sake of a quick buck (or millions of bucks in some cases). But equally disturbing were some of the stories that both Richard and I came across of what is legally done to our food. I shared these stories with anyone who would listen. In fact, some parents still run in the opposite direction when they see me at school drop-off. I soon realised that I was not the only one who was blissfully unaware of the many ways we manipulate our food.

As well as speaking with people, I read a (virtual) pile of papers in food science. I found a plethora of reports that were focused on finding more cost-effective ingredients for foods without significantly compromising the qualities of the product consumers had come to expect. These studies investigated things like mouthfeel, texture, appearance, melting point and structure, yet changes in nutritional value seemed largely ignored. Now, to be fair, I was researching fraud and looking for information regarding substitutions, but it still felt like a whole lot of jargon for looking to substitute cheaper ingredients without consumers noticing.

I'm not going to lie to you, when I began this book I thought it would be another exposé of the food industry filled with horrifying stories of how our food is manipulated. But as I got into it, my views started to shift slightly. With a better understanding of some of the processes and what drives them, things became less terrifying for me. I felt better equipped to make decisions around which processing methods aligned with my values and which didn't. It turns out I'm not terribly fond of being misled.

But before I go there, a quick definition or two to ensure we are all talking about the same thing. Food processing is any action, chemical or mechanical, that is done to food in order to change it or preserve it. This includes packaging fresh products, such as carrots, in order to ensure they reach the consumer in the intended state, namely orange and crunchy. By default, this means that processed food is the product of any of these actions. So it is in this context that I use this term throughout.

This book explores the world of food processing, telling the story through a scientific lens with the aim of removing some of the uneasiness we (myself included) tend to feel when technology is married with food. It is one of the many reasons menus tend to be devoid of scientific jargon. Citric acid conjures up images of vats of skin-melting chemicals, while lemon juice brings to mind a hot summer's day. Pyridoxal phosphate sounds somewhat intimidating, which is why we call it vitamin B_6. Alternatively, 'natural flavouring' sounds wholesome and good for the Earth. Yet, in some cases there are very complicated chemical processes that are used to extract the desired flavour compound from the natural product. Castoreum, for example, is a US Food and Drug Administration (FDA) approved natural flavour that contains an aromatic compound similar to vanilla, called acetanisole. Castoreum, however, is chemically extracted from the dried and macerated castor sac scent glands of beavers. Needless to say, it's not very widely used. Admittedly, current labelling laws don't always make it easy to be a discerning consumer, but that is another book altogether.

As consumers, we should expect high standards of our food system. We are, after all, rather dependent on it and yet the extent of global food crime and diet-related disease, as well as the secrecy within the food industry, would suggest that some aspects of the system aren't working as they should. And the timing couldn't be worse. We face a burgeoning population with rapid urbanisation and environmental uncertainty. There is such an uneven distribution of food on our planet that 15 per cent of humans are hungry while 20 per cent are obese. Food security is one of the greatest challenges our society faces for the future. There are experts around the world using a diversity of approaches to address these issues, but it is mind-bogglingly complex and it will take time that we frankly may not have.

Considering the current challenges we face with food, it is impossible not to think about how this will shape the future of food. Popular culture has painted some interesting scenarios, from the obese, leisure-suit-wearing, screen-watching, nutrition-shake-sipping, planet-killing people portrayed in the animated film *WALL-E* to the disturbingly polluted and overpopulated world in *Soylent Green* where everyone is popping green processed high-energy wafers (which turn out to be a gruesome case of adulteration, but I won't ruin it for you if you haven't seen the film). Think tanks have also generated visions of possible food futures, which range from the complete collapse of agriculture due to the loss of pollinators to 3D printed food and lab-grown meat (Chapter 5). And although this can sound like science fiction, it is already happening, but I'll get to that in Chapter 9.

But these are paralysing thoughts, and it is a time for empowerment, not debilitation! As consumers, our actions speak louder than words. We can help create the food future we envision through our purchasing decisions. To help us do that, we need to be informed and rational in our decision-making.

The ideas for this book started to form before Richard and I finished our book on food fraud. I was learning so much about how food is produced at an industrial scale and it was changing my understanding of processed food in two ways. First, when most of us refer to processed food, we generally mean highly processed or ultra-processed, 'your grandma wouldn't recognise this as food' kind of processing. But the processed-food spectrum is broad. It can mean a quick rinse and wrap of some lettuce or engineering a nano-sized (10^{-9}) emulsion in salad dressing to make it more stable ... and everywhere in between. And some foods may not sit where you would think on this spectrum. When I learned that an apple is photographed 100 times in order to calculate the blush (red on green) percentage and sort it for different retailer 'blush' preferences, I was surprised. I don't really think of a whole apple as being processed, or maybe I just thought it was a far simpler bruises/no bruises sort of affair. If I needed to know the exact blush percentage of my apple (and I don't), I do not have the tools to do this in my kitchen. Alternatively, I always thought that processed cheese slices were more of an oil-based product than dairy, yet as you will find out in Chapter 2, it was originally just cheddar that's brought to a high temperature and whisked. If I did want to make processed cheese (unlikely) I *do* have the tools to do this in my kitchen.

Second, I was reminded of the story of processed food, which provides much-needed context. In my role as the primary meal-maker, a mother and a general advocate of good wholesome food, I was only seeing the over-packaged product on the supermarket shelf. The one that contains ingredients I don't recognise and is allegedly making us all lazy and fat. Sure I see the convenience of it all, but I also feel the anonymity of it – the many unknown hands and machines that have made and delivered this product. But step back a bit and there is a very long and interesting story associated with food processing and even specific processed foods. Margarine, for example, came about as a result of shortages in edible animal fat in the 1860s. Emperor Louis Napoleon III offered a prize to anyone who could come up with a suitable substitute for butter. French chemist Hippolyte Mège-Mouriès mixed beef fat with skimmed milk and called it 'oleomargarine'. It evolved, and with each step, more of the expensive animal fat was replaced with vegetable-based fat until treating vegetable oils with hydrogen (hydrogenation) allowed animal fat to be removed altogether. During the Second World War margarine became the popular choice as it was a cheaper alternative to butter; it helped people get by during times of hardship. Then it became the healthy alternative when people were told to cut down on animal fats. And then it became the unhealthy alternative as concerns were raised regarding the trans fats margarine contains. As we learned more and circumstances changed, margarine had to evolve in order to survive in the market. Equally important, once consumers had been exposed to this easily spreadable fat in a tub, some people weren't willing to go back to hard butter. A new process of creating a spreadable vegetable-based fat had to be developed that didn't create trans fats – and it was. It's called interesterification, but let's not get into that here.

My many discussions with friends, family, acquaintances and random people on social media made me realise that I wasn't alone in having lost this context for food processing. We are focused on some negative perceptions, but should we throw out hundreds (sometimes thousands) of years of discovery because we're appalled by chemically based bacon flavouring on a triangle of fried corn dough? Surely we should at least understand the path that brought us here, to analogues for everything from crisps to meat and cheese?

It's only relatively recently that processed food has become a four-letter word – for generations before us it has helped make food more nutritionally accessible. As you'll read in Chapter 1, cutting and cooking food made it far easier for early humans to absorb certain nutrients. When our ancestors started cooking food around 1.9 million years ago, it meant that some of that time and energy that would have been needed to chew and digest raw food could instead be used for other things. The resources for developing large jaw muscles, for example, could be redirected to perhaps store a little extra body fat and maybe even grow bigger brains; time spent chewing food could be spent socialising. Then when humans began preserving foods, nutrition was improved as nutrients were available beyond the normal growing seasons and through times of hardship. Sterilisation and other processing methods have helped to remove toxins that would otherwise make food unavailable to us. Essentially, processing food (alongside agriculture) has helped our species to overcome hunger and spend more time making babies rather than chasing down the next meal.

As if contributing to our success as a species isn't a gripping enough plot, how and why different cultures evolved different processing methods is another lens through which the processed-food story can be told. Bread alone is a veritable treasure trove of anthropological insight.

An equally interesting plot is how products and methods have developed over time in response to the ever-changing sociocultural context, a shifting political and regulatory landscape and resource constraints. An outbreak can change how we handle food forever. War can limit access to resources and inspire innovation. A shift in traditional roles within the family can create a market for convenience. A natural disaster can change how we have to transport food. New scientific findings can bring to light health concerns that require us to find alternate ingredients. This information provides the much-needed context.

And last, but definitely not least, the science behind food processing offers yet another exciting storyline, and one of particular interest to me. Some of the methods and technologies developed initially for the food and beverage industry have had a significant impact in other industries, and vice versa. Louis Pasteur was as much a pioneer in reducing food spoilage as he was in preventative medicine. In his book *Chilled*, Tom Jackson gives an engaging history of refrigeration, which not only revolutionised the transportation and storage of food, but also enabled in-vitro fertilisation, the development of super-computers and even magnetic resonance imaging (MRI) scanners.

In this book, I've tried to weave these story lines together as much as possible in order to tell the story of how processed foods have evolved. Despite what we are often told, processed foods are not always the money-grabbing, addiction-forming, obesity-causing products of the big food manufacturers. Obviously some of them are some of these things, and some of them might be all of these things. But I aim to provide a different perspective by looking at how food processing, and more generally food science, gets pushed along the evolutionary path towards the latest new product. Some of these products enjoy the success of pre-sliced bread, while others go the way of Life-Savers flavoured soda (the beverage equivalent of the dodo).

Understanding this journey can perhaps help us identify if and when we started to go a bit off-track. It can help us to become discerning consumers who can identify when innovative ideas might benefit society and our planet and when they purely benefit company profits. Trained scientists and technicians are driving innovation, but we as consumers have the power to apply the brakes or sit in the passenger seat and yell 'Go! Go! Go!' And maybe we'll drive off a cliff *Thelma and Louise*-style or maybe we'll turn into the spin and get this behemoth that is the food industry on track. Time will tell. But I, for one, am not going to sit in the car and do nothing. Unless Brad Pitt is in the car, then I'll probably just stare at him.

Alongside looking at some of the major scientific advances that have helped to evolve food processing, this book looks at the earliest origins of some of the oldest processed foods on the planet: cheese (Chapter 2) and bread (Chapter 3). I investigate how methods have changed over time, particularly in this last century, and question whether these changes have anything to do with health trends, such as a rise in gluten insensitivity. From there, Chapter 4 explores how our desire to have fresh lettuce, tomatoes and mangoes all year round have driven changes in the way we process our jet-setting fruits and vegetables. I discuss how peeling and chopping up some fruits and vegetables may be encouraging better eating habits, but look at what is needed in order to keep that fresh-cut produce fresh. Chapter 5 explores one of the first processed meat products – the sausage – and how our changing attitudes towards meat have transformed this product from a method designed to use and preserve the less desirable parts into a caramelised-onion-infused treat. I try to question our attitudes towards processing methods that aim to make as much use of the animal as possible. Chapter 6 is a discussion of our love-hate relationship with salt, sugar and fat, particularly how these have infiltrated the snack and processed-food industry and what it is taking to get them back out. In Chapter 7, I look at how the family unit has changed over the last century and how this has driven a yearning for convenience that has spawned a multitude of ready-made meals and other products that put more of our nutrition in the hands of the food industry.

All things small are the topic of Chapter 8. Nanotechnology is already being used to grow, process and package our food and yet we know so little about it. What role is nanotechnology likely to play in the future of food processing, from the delivery of nutrients to smart packaging, and which of these technologies are we most likely to accept as consumers? Finally, in Chapter 9 I look at the future of food processing. It has played a significant part in our evolution and success as humans and it still has a significant role to play as we move forward into an era of environmental uncertainty and food insecurity. However, if we paint all food processing with the same brush of scepticism, we risk missing opportunities that might bring significant benefits and help us to overcome monumental challenges.

As a science writer, I love a good science story. But when it comes to my food, I can't discard my home-grown hippie roots. In my lifetime I have gone from growing everything I eat (well, to be fair, my mum grew it), to being entirely dependent on strangers for my food (save the odd potato and sugar snap pea I grow in our back garden). As a mother, I feel as though I have failed my son in not giving him the homesteader upbringing both my husband and I had. But that guilt is served with a healthy side of pragmatism. As a family, we have made choices. I spend a lot of time preparing food 'from scratch', but I also have other things I want (and need) to do. The irony that I have had to rely on more processed foods while writing this book is not lost on me. We can't afford, nor do we have the time to manage, a property for subsistence living. Nor do we have the skill, quite frankly – we would lose a lot of weight initially while we blundered through our early mistakes! And some things simply aren't practical to make myself (I refer you to the tortilla incident in Chapter 3). There are so many things to weigh up when purchasing food. As I shop, I am considering numerous factors, including food miles, human and animal welfare, packaging, sustainability, taste, quality and cost. If only the weighing scale in the produce section could produce some magic 'buy/don't buy' reading based on each consumer's values. And so, like many people, I find myself facing the convenience conundrum: which of my values (which can shift on an hourly basis, I might add) am I willing to compromise in order to make my life a little easier?

Like it or not, we have developed an extremely complex global food system. Food processing has both enabled this complexity and adapted in response to it. We can't go backwards, but we can certainly look to the past to see how drivers such as political instability, resource shortages and technological innovations have shaped the way we manipulate food. These insights might then help us to imagine how food processing, and more broadly food science, might respond to global food insecurity, health challenges and climate change to envision possible futures for food. And perhaps most importantly, it can help us to decide how we, as consumers, can help shape that future.

CHAPTER ONE

Have We Tinkered Too Much?

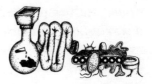

O f all the methods of food processing that I have read about, one sticks in my mind. As expected, this journey starts out with some basic recognisable ingredients. The ingredients are tipped into a macerator, where the physical and chemical transformation of the food starts. From the macerator, the pulverised food is pushed through a pipe to a mixer. Here, hydrochloric acid, water and enzymes are added and all of the contents are agitated over a period of two and a half to five hours in order to help homogenise the mixture into a paste. This thin paste then travels through a semi-permeable tube where cholic acid and chenodeoxycholic acid are added along with a cocktail of different catalysts that help speed up the chemical reactions occurring within the tube. All of the macronutrients in the paste – fats, proteins and carbohydrates – are broken into their component parts of fatty acids, amino acids and sugars, respectively. These are then diverted from the main tube for further processing and storage. The remaining gloop then travels along a slow conveyor system where millions of workers identify any remaining components of value (usually carbohydrates) and use further chemical reactions to extract them from the paste. These chemical extractions are quite complex and lead to the production of greenhouse gases, including carbon dioxide and methane. Vitamins and salts are also identified at this time and diverted for other uses. Water that has helped the processed product move through the tubes is now extracted and recycled back into the system. It takes 30 to 40 hours for the paste to travel all the way along this conveyor belt. The remaining material, no longer remotely recognisable as any of the original ingredients, is then pressed and dropped into a holding compartment until it is ready to be deposited.

By now you have no doubt recognised this as being our own digestive system, with the 'workers' being the gut flora working hard in the large intestine. However, written like this all we would need to do is wrap the end product up in plastic and slap a label on it, and it could easily pass for a description of a food-manufacturing facility. The main difference is that our pipes don't get cleaned as frequently, and arguably we have fewer quality control measures in place! And yet, when the chemical and physical transformation of food happens in factories with automated stainless steel equipment, we think of it very differently.

For me, my scepticism originates from the fact that a great deal of manufactured food isn't simply an upscaling of methods that I would use in my kitchen. Other things happen during processing that change the expected nature of the food – flour tortillas, for example, stay soft for weeks on end; or the flavour of a branded juice is consistent across seasons despite the flavour of the fruits it is supposedly squeezed from undoubtedly changing. To continue with the previous analogy, it would be like waking up one morning and finding the deposit in the toilet smelling suddenly like roses. It would be an unexpected (but not necessarily unwelcome) change in the nature of the product and it wouldn't be obvious how it was achieved.

There is an abundance of television shows, documentaries, books and articles that have tried to open up the black box that the food industry can sometimes be. Some journalists have taken enormous risks to expose animal welfare atrocities, food fraud and other issues associated with the food industry that are extremely alarming. However, other stories can be somewhat sensationalised.

In February 2016, a scandal broke when it was discovered that some pre-grated Parmesan producers were bulking out their cheese with fillers such as wood pulp. The media made it sound as though Parmesan manufacturers had a wood chipper out the back and were grinding up forests to sell as Parmesan. The wood pulp the media were referring to is cellulose – the main component of plant cell walls. It's in the leaves, the stem, the roots ... everywhere. And although we can't digest cellulose, it is an important constituent of fibre that keeps our gut healthy. Every time we eat a vegetable, we are eating cellulose. It has been used as an anti-caking agent – which stops things sticking together – in food for decades. The problem was not that they were using cellulose, but that these manufacturers were deceiving consumers by using far more than was declared and allowed in their product. It was fraud, but the focal point of the stories were around the use of cellulose, and instead of using the fibrous strands in celery to help people envision what cellulose is, the media used 'wood pulp'.

If I go back to my example of flour tortillas for a moment, I can tell two different stories about how they stay soft.

- Story one: a compound once used in anti-freeze is now used in flour tortillas (wraps) as a humectant to preserve moisture and help maintain their pliability. This compound is also used in the production of nitroglycerin, the active ingredient in explosives. It is listed on food labels as food additive E422.
- Story two: glycerol, which is a simple molecule that forms the backbone of all fats (plant or animal), is added to commercial tortilla dough because it forms strong bonds with water molecules, which helps prevent the tortillas from drying out too quickly.

Both of these stories are true, but one is clearly written to strike fear into tortilla consumers everywhere. Knowledge can help consumers sift the hype from the fact.

Of course, there is no denying that there have also been cases where the food industry has not been cautious enough in its approach and the latest and greatest food additive of today becomes tomorrow's carcinogenic substance. Just as children's toys were once painted with lead paint because we didn't know any better, decisions about what can be safely added to food have been taken based on the best available evidence at the time. And as new information becomes available, regulators have to backtrack.

A number of Sudan dyes, including Sudan I, Sudan II, Sudan III and Sudan IV were once used to add intense red colours to foods, including spices such as chilli powder (and still are, albeit fraudulently). These dyes have all been classed as group three carcinogens by the International Agency for Research on Cancer (IARC), which means that there is no evidence that they cause cancer in humans, although some studies have shown that rats have an increased risk of cancer. Many articles in the media simply say they are a listed carcinogen without giving this context. In July 2003, the European Union (EU) banned Sudans I to IV and then, in 2005, Sudan I became the centre of the UK's largest food recall to date. The UK Food Standards Agency (FSA) recalled 618 products that potentially contained imported chilli powder that contained Sudan I, which was then used to make Worcester sauce, which was then used to make a whole lot of other products, including sausages, salad dressings, pizzas, seafood sauces and an extensive list of ready-made meals. The irony, of course, is that while products potentially containing this group three carcinogen were being pulled off supermarket shelves, whole aisles remained devoted to the sale of group one listed carcinogens. These are carcinogens that have 'enough evidence to conclude that it can cause cancer in humans' – namely alcoholic beverages and processed meats, such as bacon and ham.

Other added food colours, such as tartrazine (known as yellow #5 in the US and E102 in Europe), have been linked to hyperactivity in children. Tartrazine is a yellow dye synthesised from coal tar. It's used in fizzy drinks, cereal, mustard and any number of lemon-flavoured things – from ice lollies to baked goods. The FSA has asked for the voluntary removal of these food-colouring agents, while the EU has lowered the maximum daily intake limits. Tartrazine is approved in Canada and the US, but banned in Norway and Austria – some countries are more cautious than others.

A number of sweeteners have also fallen in and out (and sometimes in again) of favour as ingredients. Aspartame came on the scene in the 1980s as the low-calorie saviour to all sugary needs. It's 200 times sweeter than sugar and therefore needs far less to achieve the same sweetness level. Then in the 1990s a report linked it to an increase in the number of brain tumours in humans, and then in 2006 and 2007 it was linked to an increase in lymphomas and leukaemias in rats. These studies were criticised and the US National Cancer Institute conducted a study of half a million people and found no link between aspartame and increased risk of these cancers. The European Food Safety Authority (EFSA) conducted a review of all of the evidence in 2013 and concluded that aspartame was indeed safe for human consumption. The safety of saccharin, the oldest artificial sweetener there is, has also been questioned. Canada banned it in 1977 when it was suspected of causing bladder cancer in rats. IARC listed it as a carcinogen in 1981, but after re-evaluating the evidence in 1999, delisted it in 2000. It took another 14 years for Canada to lift its ban on saccharin.

So, between unknown processes happening behind the very tightly sealed doors of food manufacturers, some hyped-up media, the odd bit of conflicting science, and governments implementing different regulations in response to this uncertainty, it's not terribly surprising that consumers can get a little jittery about food from time to time. As a consumer, I'm quite confident that I can make choices around the best tasting and healthiest food for my family, but an awful lot happens before I get to make that choice that I have no control over and that I have to trust others to do. This can sometimes be difficult and it is this lack of control that can cause us to throw our hands in the air sometimes and say 'what can we do about it'? And if you don't feel in control, you don't feel as though any amount of action will create change.

Directly after the European horse-meat scandal, high-street butchers' sales skyrocketed while frozen burgers and ready meals at the big retailers plummeted. Nearly half of people surveyed in the UK by a consumer watchdog organisation said that Horsegate had permanently changed their shopping habits. But less than a year later, frozen burger sales had nearly recovered entirely. The supermarkets were slashing prices, and consumers can be very forgiving when it comes to a good deal. Or perhaps we just really want to focus on the good things about food. We want to think about what to order at a favourite restaurant, or what will satisfy the latest craving, or how much that special someone is going to appreciate the amazing meal you've prepared. We don't want to think about the bacteria that are likely to be covering that chicken breast, or whether that asparagus has been coated in nanosilver to keep it fresh for longer, or whether that ground spice contains sawdust and banned dyes. Once food is in the home, a good many consumers have turned their focus to how they want to prepare and enjoy the food without thinking too much about where it came from.

The British have apparently taken this to extremes. The journalist Joanna Blythman has come to the conclusion through her extensive research that Britons don't want to eat anything that resembles living things. She eloquently refers to the ready-made meal as the 'immaculate conception' of the supermarket's chilled food section.[*]

---

[*] Joanna Blythman talks about the immaculate conception of the supermarket in her article 'Bad food Britain: Why are we scared of real food?', printed in the *Daily Mail* on 6 June 2006. [http://www.dailymail.co.uk/health/article-389321/Bad-food-Britain-Why-scared-real-food.html]

There are so many steps between this packaged meal and the farm that it is hard to even begin to contemplate where it came from, even if one wanted to. Despite the ever-popular Sunday roast, British shoppers aren't particularly keen on associating that gravy-soaked flesh with an animal that once lived and breathed. Rather than risk the smells of the butcher's shop, many shoppers prefer to purchase their meat in a supermarket, where it has been placed conveniently on a plastic tray and wrapped in a plastic force field that prevents any meat-like smells from reaching the nasal passages. The ready-made meal takes this one step further.

As well as bearing no resemblance to living things, ready meals, as we all know, are a convenience product. They can be a life-saver for some parents who have approximately 15 minutes to feed children between picking them up from after-school care and driving them to Beavers or Brownies, or for people struggling with multiple jobs to make ends meet. They have also no doubt saved many young people who have recently moved out of the family home from imminent malnutrition.

Not surprisingly, the children being raised in an era of ready meals and highly processed food are even less connected with the origins of their food. I did an exercise with seven- and eight-year-olds where I asked the children to put a series of pictures in order – from a commonly eaten processed food back to its raw ingredients. I expected them to struggle with associating a dish of bright yellow rice with the picture of ground saffron, and I certainly didn't expect them to connect saffron to a picture of a saffron crocus (though to my delight, one little girl did). I did, however, expect that all of them would know that a burger was made from mince that came from a cow. I was wrong. Some children genuinely didn't realise that mince came from an animal. They didn't connect a corn taco shell with corn on the cob either. Fish in any form other than coated sticks was just 'gross'. These are the ready-meal consumers of the future, and having been raised to it, they may not think it strange to hand over the responsibility of meal preparation to strangers and stainless-steel equipment.

So while some of us are looking for the 'immaculate conception' meal, others only feel fully satisfied if we have to scrub the soil from our carrots before peeling. We all fall somewhere along this spectrum and where we do might change from day to day. And this comes down to what factors are influencing us most that day ... even just in that moment. One day, the threat of losing your job because you didn't prepare enough for a big meeting might outweigh the risk of eating easily prepared food made by complete strangers. The next, the risk posed to our planet from industrial farming methods might have you digging your own organic potatoes from the back garden. As consumers, our perception of risk is constantly adapting to what we learn – the unknowns of the food industry, media hype, conflicting information, individual experiences and uncertainty all help to shape our perception of risk. But what we perceive as risk doesn't always match up with the real risk, and this is the gap we can close with a little more information.

So, let's start at the beginning: the reason why we first started to process food.

O ye of little face: why we process food

I watch my son playing with his Lego and he is constantly tinkering. He takes a creation apart, modifies it, tests that it does what he wants and modifies it again. He is focused as he works, persevering in his approach and innovative in his problem-solving. He is learning, which is why I can forgive him when I painfully embed a piece of his Lego into the sole of my foot.

Humans seem to be chronic tinkerers. We are continually tweaking things that function perfectly well in order to improve them, even if only a little. Our tinkering ancestors developed the first stone tools, which enabled them to slice meat and pound tubers – the first processed food. This allowed these early hominins* to chew 5 per cent less each year (yes, someone has actually estimated this). It was also some daring tinkering that led to the first thermal processing of food, which is better known as cooking. If our ancestors had never tinkered with fire, they would never have learned how to control it, and without controlled fire, there would be no cooking.

The earliest estimate of when cooking was first practised is 1.9 million years ago, although others argue that it wasn't until 500,000 years ago. There are a lot of unknowns when it comes to early humans and fire, leading to competing theories within the literature. Given this, I'm going to stick with 1.9 million years ago because it makes for the better story. But keep in mind, it is just that: a story, based on sparse fragments of evidence. Anthropologists and evolutionary biologists have arrived at this estimate of 1.9 million years ago by looking at teeth. Human molars are relatively small. In fact, in modern humans the third molars, commonly known as the wisdom teeth, may not even form or, at best, make a painful half-arsed attempt to punch through the gums. It's been a long evolution towards these wimpy molars though; they didn't just happen overnight.

Smaller teeth are first observed in *Homo habilis*, one of the earliest members of the *Homo* genus, which is thought to have lived between 2.4 million and 1.4 million years ago. Compared with earlier hominin species, *H. habilis* had a smaller face. This was probably because *habilis* was incorporating more meat into its diet and using those primitive stone tools to cut and pound that meat for easier

---

* Hominin is a relatively new term to refer to modern humans and extinct human species. Hominid, on the other hand, includes hominins plus all of the living and extinct Great Apes, such as gorillas and chimpanzees. So think of hominins as our immediate family, and hominids as the extended family that can get rowdy, break stuff and occasionally fling excrement if the mood strikes them.

chewing (remember, 5 per cent less chewing per year). This puts less strain on the chewing apparatus – the teeth, jaws and associated muscles – and the body redirects the resources for these tissues elsewhere, causing the face to get smaller.

Homo erectus, which had very similar body proportions to us and lived between 1.89 million and 143,000 years ago, also had little molars. *H. neanderthalensis,* which lived approximately 400,000 to 40,000 years ago, had relatively tiny molars despite that big sloping noggin. However, unlike *H. habilis,* the molars of both *H. erectus* and *H. neanderthalensis* are smaller than models predict they should be given their body size and skull features. This suggests something else was probably going on.

A group of researchers from Harvard decided to look at feeding time to see whether they could help explain these mysteriously small molars. Based on some basic measures – body size, mouth volume, teeth and stomach size – it's possible to estimate how much time an extinct animal was likely to have spent feeding. If they're still alive, it's much easier – you just watch them. Mountain gorillas, which feed primarily on leaves, spend 55 per cent of their waking hours putting leaves into their mouths, chewing and swallowing. Chimpanzees, our closest living relatives, have an omnivorous diet much like humans and spend about 37 per cent of their day feeding. Modern humans, because we cook and process a good portion of our food, only need to spend about 5 per cent of our day feeding. This is probably why we have Facebook and Twitter and chimps don't. In terms of primates, we (*Homo sapiens,* that is) are a significant outlier. However, it turns out that *H. erectus* and *H. neanderthalensis* were also outliers, spending a mere 6.1 per cent and 7 per cent of their day feeding, respectively. Stone tools be damned, the best way to gain this type of feeding efficiency is to start cooking your food.

Root vegetables, such as carrots and beetroot, are thought to have been a substantial part of early hominin

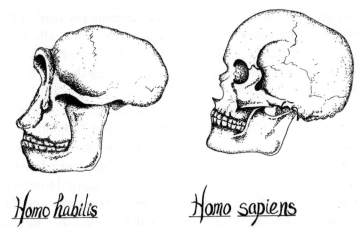

Homo habilis   Homo sapiens

Figure 1.1 Compared with the early hominin, Homo habilis, *modern humans,* Homo sapiens, *have much smaller teeth and shorter, less powerful jaws, because they process food before eating it.*

diets. Just over 2g (less than 0.07oz) of raw root vegetables takes an average of 25 chews to release 0.57kcal[*] per gram, which works out to around 20 chews per kcal. If those root vegetables are softened through cooking, though, it only takes 22 chews to release 0.56kcal per gram, bringing the chews per kcal value down to a mere 15.4. As well as reducing the number of chews, each sample of cooked vegetable requires 22 per cent less muscle to chew compared with the raw stuff. The end result is that for both meat and vegetables, cooking releases more energy for less effort. This meant that the early members of the *Homo* genus could fit a lot more into their day because they didn't have to lie around chewing all the time. But it also began an

[*] In some countries, food energy is measured in Calories (with a capital 'c'), while in others it is measured in kilocalories (kcal), but they are equivalent (1 Calorie = 1 kcal). One calorie (small 'c') is roughly equivalent to the energy needed to heat one gram of water up by one degree Celsius at one atmosphere pressure.

evolutionary trend towards big bodies with little faces, of which we are the latest iteration.

Even within the last few centuries there has been a trend towards smaller face size. Jaw size has decreased over the last century among Australian aborigines with the transition to modern, processed diets. A comparison of the skulls of sixteenth- and seventeenth-century Finnish people with modern-day Finns revealed a 6 per cent decrease in jaw length despite the overall size of the skull increasing.

Not only have we got shrinking faces and teeth, but also eating this processed food means that having a good bite is no longer a matter of life and death. As a result, the number of people with misaligned teeth or overbites or overcrowded teeth (collectively called malocclusion) is on the increase. This has often been blamed on the modernisation of our diet towards soft, processed foods. In other words, we have become such poor chewers, we have essentially become obligate food processors – we need to process our food in order to get sufficient nutrition.

It's at this point, I'm sure, that those who subscribe to a raw food diet are thinking 'what a load of hogwash'. A raw food diet, if you aren't familiar with it, is the consumption of fresh, whole plant-based foods, although some subscribers will also eat unpasteurised dairy foods and raw eggs, meat and fish. The key being that the food is, as the name suggests, unprocessed (though cutting and heating up to 118°F/47.8°C is allowed). There are many recorded benefits of such a diet, but there is also the reality that long-term raw food eaters are generally underweight and most women stop menstruating. Quite simply, they aren't getting sufficient nutrition. Given enough time, we could, as a species, potentially work our way back up to being raw food eaters, but just as our little faces and wee little teeth took a couple of million years to evolve, it would take quite a bit of time to chew our way back to some significant gnashers. Like it or not, we depend on some degree of processing.

Modern manipulations: why we process food

Not unexpectedly, the reasons we process food have changed somewhat since the days of our early ancestors. Once they slashed their chewing time with cutting and cooking, early cultures would soon have figured out that processing food could help preserve it – smoking and drying fish or meat, for example. By the time agriculture was in full swing, people are thought to have been regularly preserving food through methods such as fermentation and salting. Cultures around the world were also processing food in order to make inedible things edible. For example, in South America, indigenous cultures were peeling and crushing the roots of the cassava plant (*Manihot esculenta*), known as manioc, and then soaking it for days in water in order to remove or reduce the toxic compounds within the plant before it could be used. The early days of food processing were really about ensuring people didn't starve and didn't poison themselves – rather basic stuff.

Fast forward to the present day and food is still being processed for these same reasons, though the methods are considerably more advanced. However, beyond this, food is now also processed in order to make it far more convenient – and we're not just talking about reducing the number of chews per mouthful. In the 1920s a housewife living on a farm in the US spent just over three hours each day preparing food; her urban counterpart spent a little less, at just under three hours. By 1965, Americans were spending 2.4 hours per day preparing food, and by 2011 this number had dropped to just 30 minutes. The trend has been the same here in the UK. In 1960, about 100 minutes each day was spent preparing the family's evening meal. By 1980, this had dropped to 60 minutes and by 2013 it was down to 38 minutes. The obvious reason for this is that more women work outside the home (or, as in my case, in the home-based office). There is simply less time available for food preparation. Or is there? In Britain, the average adult still manages to squeeze in three and a half hours of

TV time per day and in America, adults are averaging five hours per day. Let's not kid ourselves; for the most part, it's a matter of where we choose to spend our time rather than having no time.

Food is processed to improve nutrition. Beyond just making some nutrients more available to us, some food processing has been used to resolve widespread deficiencies in nutrition in order to improve health and wellness. Before the 1920s, for example, iodine deficiency was common throughout parts of the US and most of Canada, causing intellectual and developmental disabilities. Governments around the world have mandated iodised salt in order to counter iodine deficiency, largely eliminating the problem in these areas. Canned fruits and vegetables are largely responsible for helping people overcome vitamin C deficiencies during winter. And by the 1920s vitamins A and D were being added to margarine to improve nutrition. Today, with the increased use of high SPF sun cream, many foods are being enriched with vitamin D as we simply can't make enough of it ourselves. Even so, just under 700 children were diagnosed with rickets, a result of vitamin D deficiency, in England in 2013/14.

Food is processed to reduce waste. Our ancestors who were subsistence farming would quite likely have been able to grow food for their needs with very little wasted. In modern food production, however, there is a greater need to process raw food materials to reduce spoilage after harvesting. In developing countries with less sophisticated storage and processing facilities, 40 per cent of food losses happen at post-harvest and processing levels. Rice, which is a major raw commodity in developing countries, needs to be dried to a minimum moisture content in order to stop bacteria and fungus growing while it is being stored ready for distribution. Many countries still dry rice in the sun, exposing it to rodents and parasites; more sophisticated processing facilities could help to significantly reduce food losses.

In industrialised nations, where these sophisticated storage and processing facilities do exist, most of the food waste is happening at the retailer and consumer stage and this is usually a matter of over-production and over-cautious consumers. Processing agricultural excesses into secondary products, such as ready meals, reduces this waste. For consumers who lack the time or skill for 'fridge management' (as it has come to be known in my family), more highly processed meals might help reduce waste at the consumer level. For example, someone might buy all the raw ingredients to make a homemade lasagne. They only use a portion of the fresh produce they buy to make the lasagne and an accompanying side salad. They then need to set aside the time to prepare other meals that week with the remaining fresh produce, otherwise it will get wasted. Had they bought a ready-made lasagne and salad bag to begin with, there would probably have been no waste. In other words, it's not just about taking the time to prepare meals from scratch, it's about having the time to manage the continuously degrading ingredients as well. I admit that sometimes it feels as though there is a time bomb on the refrigerator door and every time I open it to see something looking wilted I feel the pressure of having to deal with it right then and there. It can be stressful!

Another very modern reason for food processing stems from the strange and largely untrue belief held by retailers in industrialised nations that consumers can't handle ugly produce – processing for aesthetics. This, to me, is an enormous tragedy. Not only because when surveyed, most consumers have said that so long as taste isn't compromised, they aren't particularly fussed if their carrot looks like an astronaut (search for it online if you haven't seen it already, it's very funny). Yet, UK food waste campaigner Tristram Stuart found that between 12 and 15 per cent of carrots grown by M.H. Poskitt Carrots in Yorkshire were out-graded because they didn't look right – they were either bent, insufficiently orange or had clefts or blemishes. They were being redirected into animal feed, but as they are perfectly fine for human consumption, further processing into soups or other products could keep these ugly little fellows within the human food chain and potentially be worth a lot more to the farmers who grew them. Luckily, more supermarkets seem to be embracing the 'love the ugly' movement and are offering less than perfect produce at a discounted price. Woolworth's in Australia ran the 'Odd Bunch' campaign and ASDA in the UK sold fruit and vegetables that were 'Beautiful on the Inside', and similar campaigns have run in the US and Canada. France retailer Intermarché took this one step further, following the success of its ugly fruit and vegetable campaign, and started selling ugly cakes and cookies as well.

Food is processed to provide diversity across seasons. While we might be satisfied to eat ugly food, we are less excited about surviving on kale and cabbage through the winter. Overwhelmed by choice, we are no longer satisfied with eating only frozen, canned or otherwise preserved produce in the off-season. We want to have fresh products available all year round. As a result, our jet-setting food now travels further between where it is grown and where it is sold than ever before. Out of this comes a whole area of food processing that is involved in transporting fresh produce and extending its shelf life for as long as possible. Cut salad is an excellent example of this. Salad picked from the garden simply needs to be washed and torn into bite-sized chunks. In order to provide a convenient bag of salad with different varieties of leaves throughout the year, quite a bit more processing is required.

If you've ever grown your own salad mix you will know that most varieties, particularly the tender baby leaves that are so popular, start to wilt as you walk back from the garden towards the kitchen. This is partly because a cut leaf immediately starts to respire faster; chopped kale, for example, has an 88 per cent increase in respiration rate compared with a whole kale leaf. The leaf is quickly losing water through respiration and it is also using up its stored sugars, causing it to wilt. To add to this, enzymes are released from the cut or torn area of the leaf, and these immediately start to act on the surrounding tissues and cause browning.

This presents a significant challenge to the food industry: leaves are wilting faster than they can be packaged. The fresh-cut leaves are washed about three times, usually using a chlorine solution or ozone, in order to eliminate microbial growth. The leaves are then dried before being packaged in some very fancy packaging. The bags have a patented film inside that controls the amount of oxygen within the package. Reducing the oxygen (and increasing the carbon dioxide) reduces the respiration of the leaves, but if oxygen is eliminated completely, anaerobic respiration happens and the lettuce spoils even faster. It's a fine balance, but this modified-atmosphere packaging can increase the shelf life of loose lettuce leaves to 15–17 days. Of course, alternatively consumers could just buy a head of lettuce as it will keep a lot longer without all of the fancy packaging, but more thoughts on that in Chapter 4.

Jet-setting food isn't just about giving consumers choices, though. Advances in the storage, packaging and transport of food have enabled the industry to source cheaper labour for processing as well as to strategically place processing in countries with less stringent environmental regulations. It's also enabled the development of exports that wouldn't otherwise be possible. The result is the situation currently happening in east Africa. Countries like Ethiopia are exporting goods worth £690 million ($900 million*) a year in coffee, £440 million ($570 million) in vegetables and just over £460 million ($600 million) a year in cut flowers, while their own people are experiencing extreme famine. There is something very wrong there.

---

* US dollars, unless otherwise stated.

Processing for profit is another modern motivation. And for this I can think of no better example than snack foods. Pretzels and popcorn are often regarded as the oldest snack foods; an ancestor to the modern pretzel is thought to have been first baked in Italy in the sixth century, while popcorn is thought to date back over 5,000 years, having been first used by the indigenous people of North America. Both of these things are recognisable as food – a pretzel is no more than a modified bread and popcorn is simply an explosive variety of corn. The snack foods that fill the aisles of supermarkets and convenience stores everywhere today, however, are far less recognisable as food. At their base is usually some form of corn, rice, potato or wheat. From there, however, an arsenal of processing techniques are used to make these basic ingredients unique by extruding, puffing, crisping or baking them into a diversity of shapes and textures. They are then coated in top-secret combinations of flavourings, salt, fat and, in some cases, sugar, that make them thoroughly moreish. Manufacturers of these highly processed products aren't interested in providing a nutritional product; they are interested in making money out of consumers with a snack-mentality, who crave quick empty calories. Americans consumed 9.5kg of savoury snacks per person in 2015 and in Britain consumers managed marginally less at 7kg each. The savoury snack market was valued at £71 billion ($94.5 billion) in 2016 and is expected to reach £103 billion ($138 billion) by 2020 – a phenomenal projected growth. There is no question that these are highly processed; the question really is whether most of these snacks (except popcorn and pretzels of course) still constitute food.

Companies have also found profit in added-value products. Whole carrots, for example, sell in Tesco for £0.45 per kg, but if they are peeled and cut up into batons, they sell for £1.67 per kg – more than three times more. In the US and Canada, consumers are probably paying three times more for the very popular peeled and shaped carrots known as baby carrots. Of course, I could have just as easily put this under the paragraph about convenience, but consumers started to buy these convenient products because they were available. Supermarkets and producers put them there because they could see the cash cow called 'convenience'. It wasn't as though consumers had stopped eating carrots because they were too much work! Well, actually, that's not entirely true, but more about that in Chapter 4.

And finally, we come back to the human propensity to tinker because some food processing seems motivated simply by a desire to innovate (which is really just tinkering with purpose). A significant amount of innovation is driven by commercial interests, with R&D departments working alongside marketing departments to find the next processed product that will shake up the industry and bring sizeable profits. It might be as revolutionary as pre-washed bagged salad or just a slight modification of an existing winner, such as a new flavour of crisp. Some of these innovations seem intended to push the envelope and test consumer limits – vegetable-flavoured jelly (Jell-O) of the 1960s and purple ketchup of the early 2000s come to mind – while other innovations seem as though they are simply the product of overworked and underpaid post-docs suffering from a craving as they work through the night – 'I wonder what would happen if I put melted chocolate in the 3D printer?' As expected, many of these innovations die a natural death (celery jelly was never a big hit), but some may go on to broaden the culinary horizons of future astronauts or Mars colonists, who knows? With our basic nutritional needs mostly met, it seems natural to play with our food a little and learn from those experiences – it's like standing on a piece of Lego, we tolerate the purple ketchup on the off-chance that it somehow helps advance food science.

Transforming tucker: how we process food
Although the specific technologies used have become more sophisticated and automated over time, many of the basic methods of processing food have remained relatively

unchanged; at its core, processing is still mainly cutting, grinding, mixing, heating, cooling, drying, fermenting and packaging.

Heating food remains one of the best ways of killing off unwanted pathogens and extending the shelf life of food. Those first cooking hominins no doubt realised that as well as being easier to chew and (in some cases) tasting better, some of the foods they cooked also lasted longer. Outside of cooking, there are three main ways of thermally treating food: canning, pasteurising and blanching. These are all things that many of us have done in our own kitchens.

Canning began in the eighteenth century when the French chef Nicolas Appert began bottling up soups, vegetables, jams and meats in thick wide-mouthed glass bottles in his own home experiments. He left a gap at the top of the bottle, shoved a cork in, and then wrapped the whole thing up in canvas before boiling the heck out of it. This was, for all intents and purposes, sterilisation as it probably killed all living things contained within the jar to stop the food from spoiling, but it also no doubt changed the nutrient content, taste and texture of the food at the same time. Needless to say, it wasn't yet an exact science and Appert determined the appropriate sterilisation times for different foods largely based on trial and error.

In 1795, the French government launched an award of 12,000 francs to anyone who could develop a new method of preserving food – Napoleon's Food Preservation Prize. During the Seven Years War, France and Britain had lost half of their seamen to malnutrition and things weren't looking much better for French seamen or soldiers in the Napoleonic Wars. Salted meat wasn't cutting it in terms of providing a variety of nutrients. After many years of experimentation, Appert submitted his invention in 1810 and was given the award on the condition that he make his methods available to the public. He did so in the form of the book, *L'Art de Conserver, Pendant Plusieurs Années, Toutes Les Substances Animales et Végétales.*

Philippe de Girard, an inventor from France, decided to take advantage of England's entrepreneurial spirit. With Appert's methods fully open to the public, Girard bottled up foods and brought them over to the Royal Society to taste. In 1811, a patent for the method was awarded in the name of Peter Durand, a British merchant, as it was hardly appropriate for a Frenchman to hold a patent in England at the time. The patent included tin cans among the preserving vessels listed, which was a development from Appert's methods. Durand sold the patent on to an engineer by the name of Bryan Donkin for £1,000, who then commercialised the idea. He refined the methods and built a factory close to the docks on the River Thames and the first tinned food rolled out through the factory doors in the summer of 1813. The tinned food was tried out on various members of the royal household, who praised it highly. It was then sent to British sailors and soldiers, who were no doubt relieved to eat anything that wasn't salted meat. By 1821, the Office of the Admiralty and Marine Affairs was ordering 4,082kg (9,000lb) of tinned food per year for the British Royal Navy. It would be another hundred years, though, before canned goods would become cheap enough for the average family to buy. These days you'd be hard pressed to find a household without some canned goods in the cupboard.

The idea of pasteurisation came from Appert's finding that heating prevented food from spoiling. Unlike sterilisation, however, pasteurisation is usually applied to liquids and involves heating the food just enough to slow down the growth of most of the microorganisms without killing them all. In the late 1850s, the French scientist Louis Pasteur was experimenting with fermentation and had identified yeasts as the organisms responsible for alcoholic fermentation. In 1859, Pasteur disproved the widely held belief of the time that the microorganisms involved in fermentation just spontaneously appeared in the liquid. People thought that animals suddenly materialised – maggots grew from decaying meat and mice from sweaty underpants. Pasteur proved through a series of experiments that microorganisms only grew in sterilised liquid if that liquid came in contact with dust particles in the air – in other words, the liquid had to be contaminated. From there, Pasteur showed that microorganisms were responsible for spoiling beer, wine and milk, but that the growth of these organisms could be slowed through gentle heating. By the late nineteenth century, milk and beer were being regularly pasteurised. Today, milk, nuts, vinegars and juice are generally pasteurised.

Blanching is an even milder form of thermal processing as it is mainly intended to slow the activity of enzymes that would otherwise change the flavour, texture and colour of the food. It involves a two- to three-minute dip in boiling water followed by rapid cooling in order to stop the cooking process, and it's usually applied to vegetables and nuts. Blanching also softens the tissues and reduces the number of microbes on the surface of the food. We always used to blanch our vegetables before freezing them on the farm. At an industrial scale, blanching is usually done using a hot-water bath or with steam. In the last decade, microwave blanching has also been introduced commercially.

On the opposite end of the spectrum from heating, there is refrigeration and freezing. In 2012, the Royal Society named refrigeration as the most significant invention in the history of food and drink. Cold temperatures extend the shelf life of food not by killing microbes, but by making it too cold for them to thrive and grow. And, like blanching, refrigeration slows the activity of those pesky enzymes that cause fruit and vegetables to rot.

The benefits of keeping food cool were realised long before the refrigerator had been invented. There is evidence from the Stone Age (9000 to 2000 BC) that pits dug into the ground, where a more constant temperature could be achieved, were used to store food. There is also evidence from China, Greece and across the Roman Empire that pits were dug close to natural sources of ice, such as lakes or ponds, so that the ice could be placed in the pits with the food in order to keep it cool. However, it was the ancient Persians who mastered the ice house in around 400 BC in the deserts of Iran. They built huge conical mud brick structures, called yakhchāls, rising about 20m (60ft) high. On the north side of the yakhchāls, shallow pools were constructed behind walls built to keep them shaded for most of the day and protect them from wind. Aquaducts would bring water to these pools and on winter nights, ice would form. This ice would be harvested before sunrise, brought down into a pit in the ice house, and layered between insulating straw. The deep pit took advantage of the regulating temperature of the earth, and the cone shape above ground channelled hot air up and out, much like the vents of a termite mound. Once the house was filled with ice, it would be sealed up and then reopened in summer so that the ice could be used to keep things cool in the desert. By the seventeenth century, ice houses were quite common on larger estates in Britain. These structures were not nearly so magnificent as the Persian versions, being very simple domed buildings made of brick or stone. Examples from the nineteenth century are still around today, including one not far from where I live, on the Ashton Court estate in Bristol. These ice houses certainly weren't terribly convenient for midnight fridge-raids, but they served their purpose.

Prior to refrigerators or electricity reaching the outback, Australians had the Coolgardie safe, which was invented in 1890 and worked on the premise of evaporative cooling. It was a relatively small appliance with a timber or metal-framed cabinet with open sides, and it was covered in hessian (burlap) fabric. On top there was a tank filled with water, which had felt strips that wicked up the water from the tank and carried it down to the hessian fabric. As the breeze blew past the Coolgardie, which was usually positioned on a veranda, the water in the fabric evaporated, absorbing heat from the cabinet and keeping the contents within it cool.

In the US and Britain, at the beginning of the twentieth century, many households had iceboxes, which weren't much more than insulated cupboards that were pushed up along an outside wall of the house. A hole was cut in the wall so that the ice man could simply open a small door and slip a 25–50lb block of ice into the back of the cupboard, without having to come into the home. It wasn't until 1913 that the first refrigerators were built for home use and not until after the Second World War that they would become popular in the average British or American home. Now, modern fridges are slowing the growth of microbes and the degradation of fresh food everywhere. They can't stop the deterioration of food indefinitely, though, and there are many experiments being carried out inadvertently in the back corners of fridges everywhere as evidence of this. Modern fridges can even pour water and make ice right in the door – the ancient Persians would have been mesmerised!

Freezing, of course, can extend the life of food even more than refrigeration – frozen food can easily last a year or so without compromising quality. However, there is a small cost to taking food to these lower temperatures. Freezing essentially stops microbial growth completely, but as all living materials contain water, ice crystals form within the tissues of the food, which can sometimes have an impact on the texture and flavour. Commercial freezing began in the late nineteenth century to preserve foods such as meat, butter and fish during transport. But in 1928, American inventor Clarence Birdseye revealed a quick freezing method, called a double-belt contact freezer, which could freeze fish and vegetables so quickly that very few ice crystals formed. This is when frozen-food production became commercially viable, but it didn't really enter consumer homes to any significant level until the 1940s, when separate compartment freezers became available. This was good timing as tin was becoming scarce in the early 1940s, since the world's biggest supplier of tin, China, was trying to defend itself against Japanese invasion. This limited the production of canned goods. Frozen food, which was wrapped in paper and cellophane, became the better option. Later that decade, the first frozen pizza was launched and frozen-food sales have been growing ever since. In 2016, they were valued at about $53 billion in the US and about £4 billion in the UK. Americans in particular have had a long love affair with frozen ready meals, leading some retailers to specialise entirely in frozen foods.

These days there are numerous ways foods are flash frozen, from gigantic chambers that blast cold air at the food, to baths of liquid nitrogen. It is widely considered to be the best method of food preservation in terms of maintaining the taste, texture and nutritional value, and yet there remains a bit of a stigma among some consumers when it comes to frozen food. It's often cheaper than fresh, so therefore it must be inferior, right? Not necessarily. First of all, vegetables and fruit destined for flash freezing are picked when they're ripe, compared with fresh produce that is generally picked underripe and artificially ripened closer to the point of retail. Frozen produce is also usually frozen within a few hours of being harvested. If you consider that the produce starts to degrade the moment it is picked, a few hours is nothing compared with 'fresh' beans and baby corn that have been a week travelling from Kenya or asparagus that's come from Peru.

The quality of fresh versus frozen is still highly debated and might be more a matter of personal choice than anything. However, where there is little room for argument is in how frozen food can help combat food waste. Frozen food is cheaper for the consumer not because it is somehow inferior, but because producers and retailers don't have to factor in a certain amount of loss due to waste before it even gets to the consumer.

In Britain, and indeed in most developed countries, the majority of food waste happens at the point of retail and with the consumers. A study conducted by Sheffield Hallam University, UK, in 2014 showed that Britain could slash its food waste by almost 50 per cent if more families incorporated frozen foods into their regular meal plans. This would be an enormous dent in the estimated 7.3 million tonnes of avoidable food waste generated in UK households each year! Frozen food generates 47 per cent less food waste compared with fresh foods because people can portion out some frozen carrots for a meal and put the rest back in the freezer, rather than discovering some black slimy things a couple of weeks later in the (not-so) crisper drawer.

Another method of extending the shelf life of food that doesn't involve heating or cooling is irradiation, which was established as a scientifically sound and safe method of processing food by the mid-1900s. It isn't widely used commercially, however, because consumers generally aren't keen to see the words 'irradiated' and 'food' in the same sentence – there is a perceived risk, whether it is justified or not. Pretty much every government website on the subject of irradiated foods starts with the sentence 'irradiation does not make food radioactive'. They are clearly aware of the misconceptions out there. The confusion comes from the fact that a radioactive material, usually cobalt or caesium, is used in the process.

This is how it works. Food (packaged or not) is placed in a room with artificially produced cobalt (or caesium), which has an extra non-charged particle in its nucleus – a neutron. Much like the obnoxious uninvited dinner guest, this additional neutron makes things very unsettled, unstable even, in the nucleus of the atom. So, in order to restore some order, this extra neutron splits into two charged particles – a proton and an electron; the proton stays in the nucleus, while the electron is ejected out of the nucleus. (If only we could do that with the annoying dinner guest.) This is radioactive decay. The cobalt has decayed into a stable, though slightly excited, nickel atom, which then calms itself down by releasing a couple of parcels of energy (gamma rays) that radiate out from the atom. It is this radiation that interacts with the food, just as thermal radiation produced in your oven makes molecular changes in your food that cause it to brown during cooking, or microwave radiation produced in a microwave oven agitates water molecules in food to generate heat. Gamma radiation knocks away any weak electrons in the food, which stops the activity in living cells, such as parasites or insects, that might be living on or in the food. It also prevents potatoes and onions from sprouting, helping them to keep longer. The radiation is too weak to make anything radioactive, but just strong enough to cause a few changes that kill off unwanted pests.

Most countries have a list of food that is allowed to be irradiated along with limits on the radiation dose. They also require that irradiated foods be clearly labelled, often with the radura logo (see figure) clearly displayed. For mixtures of food where only some of the ingredients have been irradiated, it can get more complicated. In Europe, any foods that contain irradiated ingredients should have 'irradiated' or 'treated with ionising radiation' clearly marked beside that ingredient on the list of ingredients. In Canada, if irradiated ingredients make up 10 per cent or more of the food, they must be labelled as 'irradiated' in the ingredient list. In the US, irradiated ingredients in multi-ingredient foods do not need to be labelled. The most commonly irradiated foods include potatoes, onions, wheat, flour, whole or ground spices, and fresh and frozen raw beef mince.

Opponents of food irradiation give a number of reasons for avoiding it and once we eliminate the fallacy of greengrocers needing to wear hazmat suits, there are three main arguments.

- First, food irradiation kills all bacteria, including the friendly ones. This isn't true; the dosage is controlled so that it is essentially the equivalent of pasteurisation (though obviously some argue against pasteurisation as well).

- Second, irradiation damages the nutrients in our food. Well, it could be argued that every processing method damages at least some of the nutrients. We forget that even air and light (natural and artificial) will damage the nutrients in our food. For example, 75 per cent of the riboflavin (vitamin B2) is lost from milk packaged in a clear container after just one hour of exposure to sunlight.
- Third, irradiation produces free radicals that can harm us. Free radicals are indeed produced by irradiation and most of these decay long before the food reaches our mouths. However, some do persist. Yet in one of the most commonly irradiated foods, dried herbs, it's impossible to tell the free radicals from irradiation apart from free radicals that naturally occur as a result of the decaying process. Free radicals are everywhere – in a slice of cake or a glass of wine – and far more are produced in our own bodies through sun exposure or stress than by eating irradiated food.

There isn't even a radioactive waste problem. The radioactive material that is used to irradiate food is produced by exposing non-radioactive cobalt or caesium to intense radiation. When it has decayed below a certain threshold, it is simply returned to the facility for 'recharging'. In other words, it's recycled. I'm not an advocate for irradiated food, but I think it's important to make decisions based on fact rather than fiction.

Drying and smoking are traditional methods of dehydrating foodstuffs that also help to make them last longer as they not only slow enzymatic activity, they also create a less desirable environment for moist-loving moulds. There is no doubt that dehydrating food completely

Figure 1.2 The radura logo indicates that food has been irradiated.

changes its characteristics, but this isn't necessarily a bad thing. By removing most of the water, the food becomes significantly lighter and potentially more transportable, opening up a world of 'instant' convenience foods from coffee to soup.

Smoking not only dehydrates the food, it coats food in a complexity of flavours that give the sensation of a campfire in the mouth – well, that's if you burn only apple wood or hickory on your campfire. The chemicals in the smoke not only give flavour, they also kill some of the bacteria found on the surface of the food and slow the growth of others. Some of the chemical compounds even stop the oxidation of the food, slowing its deterioration. Of course, these days very few producers can be bothered to string fish up by the fire and instead turn to liquid smoke, but this only provides flavour without any of the preserving qualities.

Like cooking, dehydration has been used to preserve foodstuffs for a very long time, and for some foods these methods haven't changed much at all. Rice is often still dried using traditional systems: it is sun-dried on mats or canvas or, in the case of larger-scale production, large concrete pads. Many herbs and spices are also still dried using traditional methods. These systems are straightforward and inexpensive, but they are also at the mercy of the elements as well as rodents, insects and other pests that will gladly take the opportunity for an easy meal. More modern methods provide complete control over the process and obviously eliminate the little beasties, but they are also incredibly energy intensive; research is under way to improve that.

Fermentation is yet another ancient method of preserving food that continues today and has become particularly trendy again in recent years. Honestly, what would we do without microorganisms? Considering all of the things that we hold international days of recognition for, surely there should be an International Microbiome Day to appreciate all those microbes working hard in our guts. There could be a probiotic food court, a methane maze challenge (smell your way out) and free stool sampling! OK, maybe not. The fermentation of food that happens in our gut and provides us with an estimated 10 to 20 per cent of our nutrition can also happen outside the gut, converting perishable products into new culinary delights. We are familiar with beer, sauerkraut and yogurt as fermented foods, but cheese, salami, soy sauce, Tabasco sauce and cocoa are also products of fermentation. There are examples of fermented products from every country in the world. Kefir (fermented milk), kimchi (fermented vegetables, mainly cabbage) and kombucha (fermented tea) have all increased in popularity in Europe and North America lately, with increasing attention to maintaining a healthy gut community through eating fermented foods.

When we ferment food, we are encouraging the growth of desirable, non-illness-causing bacteria, which change the chemical composition of the food enough that the nasty microbes can no longer grow there. Sometimes we help them along in their task, for instance in creating the acidic environment in which pickled onions or cucumbers are fermented (pickling is just another form of fermentation). These beneficial microbes produce natural preservatives through their fermentation processes, including lactic acid and ethanol, which make the food a less desirable place for other microbes to grow. They also produce carbon dioxide, which reduces the amount of degrading oxygen in the food.

Some food processing includes a fermentation step to encourage the production of one or more of these natural preservatives, though perhaps not entirely for the purpose of preserving. Yeasts are incorporated into bread dough and encouraged to ferment the sugars to create carbon dioxide that expands the structure of the bread. Yeasts are also encouraged to ferment sugar and produce ethanol in wine, beer and other alcoholic beverages. We owe a lot to these single-celled wonders!

Among the other greatest inventions in the history of food and beverage processing, the Royal Society listed baking, which differs from cooking in that the heat changes the structure of the product, which is usually flour-based. Grinding and milling and the microwave oven also made their list of great inventions. And in terms of 'packaging', the barrel and cork were seen as revolutionary, yet are nothing compared with today's modified atmosphere packaging.

Industry magazines, such as *Food Engineering*, list other advancements in food processing as being pivotal. These advancements aren't necessarily based on their impact within the scientific community and society more generally, but rather on their ability to aid the industry to meet productivity demands and food safety standards, and to improve profit margins. These include sliced bread and cellophane, which both came about in 1930, and the use of ethylene to ripen fruits, such as bananas and citrus, which began in 1936. The development of the first artificial sweetener, saccharin, in 1879 is also a critical moment in food manufacturing history. Artificial sweeteners are regulated as food additives. Gram for gram they are usually the same number of calories as sugar (sucrose), but they are many, many times sweeter, so far less is needed. For consumers, artificial sweeteners mean fewer calories in their favourite foods and potentially fewer visits to the dentist as sweeteners don't stick to tooth enamel like sugars do. But it's not just about calories and cavities; these sweeteners also avoid creating glucose spikes that are extremely dangerous for diabetics or people with other metabolic disorders who are affected by high blood glucose levels. That is, if you can get past the taste they leave in your mouth (I can't). For the industry, these artificial sweeteners mean lower costs because less needs to be used, and they can be stored indefinitely without risk of spoilage. The problem, however, is that they don't always perform the same way in the food. In other words, an artificial sweetener has a different feel in the mouth and changes some of the properties of the food when it replaces sugar. A low-calorie ice cream, for example, with artificial sweeteners, also has carrageenan (which comes from red seaweed) added in order to give it the same viscosity and feel as its higher-calorie sugar-added counterpart.

Food additives have certainly helped to propel food processing; the carrageenan in the ice cream is only one of hundreds of food additives used by the industry. And while it is argued that food additives have been used for hundreds if not thousands of years, with salted fish and meat held up as one of the oldest examples, it is within the last century that the world of food additives has truly exploded. The line between additive and ingredient is a bit blurry, but generally food additives are ingredients that are added in very small quantities (usually less than 2 per cent of the overall weight of the foodstuff) to create a technical effect. Examples include leaveners, stabilisers, anticaking agents, colours, flavours, emulsifiers and preservatives. As analytical chemists and food scientists gained a better understanding of chemical compounds and their properties, it became possible to specifically engineer compounds that would have very specific effects. Before that, a more trial-and-error approach led to some very interesting processing agents and additives, with some undesired effects. For example, in the nineteenth century, in order to achieve pickled cucumbers with an unnatural and yet attractive green colour, people would either heat the pickling solution in a copper kettle or with a few copper coins or blue stone, which is copper sulphate. Bright green pickles were indeed the result, but of course, copper is toxic in greater doses, putting pickle fans into peril (say that 10 times fast). People simply didn't know that. These days, food additives undergo far more rigorous testing before being approved for use in food, but that doesn't mean a few haven't lost their approved status, Sudan dyes being a good example.

One last processing method I should mention is extrusion, because it is so commonly used in food manufacturing. As you can guess by the name, extrusion involves pushing a food mixture through a specifically engineered opening to create a desired shape. The food is usually brought up to a high temperature for a short period of time during extrusion and it is through this combination of heat, pressure and shearing forces as a piston or screw forces the mixture through an opening, that the desired shape, texture, functionality and flavours of the food can be produced. Commonly extruded foods include pasta, noodles, breakfast cereals, flat bread, snacks such as Wotsits (Cheetos), corn curls, crisps, crackers, chewing gum, chocolate and chewy sweets.

And there you have it – the general ways in which we chemically and physically alter food these days. We have been tinkering with these methods for a very long time, but within the last century we have applied scientific rigour to diversify and hone processing methods in order to produce more food, more efficiently and safely, and with greater convenience to the customer. Things are more controlled, mechanised and sanitised than ever before. Food isn't just frozen; liquid nitrogen is used to quickly cool individual corn kernels, peas, berries and other small items down to -196°C nearly instantaneously so that ice crystals don't even have time to form. Things aren't just packaged; they are wrapped in a specifically controlled atmosphere in order to reduce spoilage. Scientifically, it's rather incredible. And yet, it does feel as though in some cases, we've gone too far.

This is a question I have asked myself often over the last few years and I am constantly trying to define where my comfort limits lie. Admittedly, these have changed as I have learned more about food. I am not necessarily distressed by the processes themselves. I once made fresh pasta myself at home without the convenience of a pasta maker and I'm not going to lie to you, sweat and tears were involved. I didn't think it was worth it. I am perfectly happy for a big stainless steel machine to push out my pasta. Extrude away, I say! Yet, I would never buy pre-packaged pancakes because they are so simple to make at home and because the store-bought ones contain preservatives. I would much rather make a big batch at home without the preservatives, throw the extras in the freezer and toast them individually over the next month or so as I want them.

So is it the additives that are so much a part of modern processing that I'm not keen on? Well, yes and no. The store-bought pancakes in my example above are enriched with calcium, iron and vitamins B1 and B3, which makes them arguably more nutritious than my homemade pancakes. Additives here seem to be a good thing. However, I would prefer to find those nutrients elsewhere in my diet rather than have them with a side of preservatives. But that's because pancakes are easy to make. In contrast, I buy prepared mustard, which also contains preservatives, but I am willing to overlook this because, quite frankly, I can't be bothered to whip up mustard every time I want to make a sandwich. I am quickly knocked from my high horse by convenience.

I get irked by food additives that seem to do nothing more than mask and deceive the consumer. Colourings that give consistency or mask natural variation in ingredients or cover up inferior ingredients, for example. Additives that make things appear thicker, more homogeneous or more flavourful all seem somewhat dishonest to me. However, discerning the good and bad additives on any ingredients list (and I add that these categories are completely subjective) requires nothing short of a chemistry degree.

Is it the automation and scale of it all? As I said in the introduction, I don't like the anonymity behind industrial food production and processing – if given a choice, I will always choose the food with the better story. But as I explained with the pasta, there are many cases where I'm perfectly happy to hand production over to complete strangers for the sake of convenience. Scale is often tricky too. There are definitely some efficiency gains to be made with increased production of certain foods, but I absolutely want to support small, local production and will continue to do so when I can afford to.

I have come to the conclusion that for the most part, when it comes to processed food, I find that I draw the line when the processing methods have changed the nature of the food beyond what my expectations are. So, for example, a loaf of bread stays fresh for at least twice as long as my homemade loaf, due to additives. Or my store-bought 'freshly squeezed' orange juice doesn't separate in the fridge like home-squeezed does. Or my store-bought pasta sauce in a jar is far brighter and sweeter than my home-canned pasta sauce. Of course, being able to draw the line in terms of what is acceptable in food processing first requires having realistic expectations of what processed foods should do. We are losing some of that knowledge in the modern kitchen. Many people have no idea that it is not natural for a loaf of bread to stay fresh for a whole week on the counter or that freshly squeezed orange juice naturally separates. This is where we can take a little responsibility as consumers to educate ourselves. And then we can decide for ourselves where we have perhaps tinkered too much with our food and pushed it beyond what is necessary or even beyond the definition of food.

CHAPTER TWO

Maturity Doesn't Necessarily Come with Age

Cheese, like wine, is one of those food items on which most people have an opinion. There are cheese enthusiasts who throw caution to the wind in favour of a memorable cheese experience. My husband, much to my chagrin, is of this ilk and believes that cheese should be capable of crawling onto the cracker of its own accord. And some do. The traditional Sardinian cheese *casu marzu*, for example, is alive with the writhing little maggots (larvae) of the cheese fly (*Piophila casei*). Cheesemakers encourage the adult female flies to lay their eggs in the hard ewe's milk cheese (pecorino) by cutting a hole in the top of the round and leaving it outside. When the maggots hatch, they begin to munch on the cheese and in the process break down the fat, converting it into a smoother, softer product as it passes through their digestive tracts. The result is a mixture of crumbly undigested pecorino and soft larvae excrement that is best smeared on bread. To add to the experience, the larvae have acquired the name 'cheese skippers' because their flight response is to convulse their body with enough vigour to launch them four to five inches in the air, away from imminent danger. So as you bring the cheese to your mouth, you are rewarded by the sight of panicked larvae leaping for their lives.

To date we have managed to avoid wrigglers in our cheese at home, but the cheese compartment in our fridge is nonetheless a living experiment of cross-contamination and over-ripening. My husband has indeed managed to turn a small block of mozzarella into a Brie by wrapping them together. The stinkier the better is the motto at our house! And I find myself compelled to explain this to all guests who enter the kitchen when the fridge has been recently opened.

At the other end of the spectrum are those who prioritise safety (and probably price) over experience. Venture into the cheese compartment of their fridge and you are likely to find a big block of mild or medium cheddar, Monterey-Jack or mozzarella (pre-grated or block, never balls). They will probably be supermarket own-brand labels and if you're in North America, the cheddar will probably be orange. I am not passing judgement. After all, these are people who can open their fridge with abandon without house guests fleeing. They are investing in a consistent, reliable, low-return product – the savings bond of cheese if you will.

Luckily, with over 700 different cheeses produced in the UK alone (according to the British Cheese Board), there is something for every palate, regardless of where you fall on the cheese spectrum. Brits eat just under 12kg (about 26lb) of cheese per person per year. We are lightweights compared with other European countries, though. In France, the per capita consumption of cheese is nearly 27kg (59.5lb) per year and Finland and Germany are close behind with about 25kg (55.1lb). That's more than my nine-year-old son weighs! In North America, Americans are munching down about 15kg (33lb) of cheese each per year and Canadians just less at 12kg (26lb), similar to the British, while Mexicans eat a mere 4kg (8.8lb) each per year. And in China, people are eating less than 100 grams (3½ ounces) each per year.

In 2012, the global cheese market was worth £61 billion ($79.6 billion) and it is estimated it will be worth over £77 billion ($100 billion) by 2019. It's big business. The cost of individual products can range from a run-of-the-mill supermarket cheddar costing about £5.50 per kg ($14.81 per lb), to very specialised products, such as moose cheese (£169 per kg or $455 per lb) or Serbian pule, made from

Balkan donkey milk, which sells for as much as £734 per kg or $1,976 per lb. But to truly understand the value of cheese, it's necessary to look back to its origins.

The early curd hypothesis

The history of cheese making begins with our innovative yet lactose-intolerant Neolithic ancestors, who recognised the nutritional value of milk. Approximately 10,000 years ago people began trading in a hunter-gatherer lifestyle to become herders and farmers, domesticating various plants and animals. While different populations around the world would discover agriculture at different times during this period, it was the Fertile Crescent, which runs from the Persian Gulf to the Mediterranean, that is credited with kick-starting this transition to an agricultural lifestyle.

Bone shards recovered from archaeological digs in south-west Asia suggest that these cultures began domesticating goats and sheep around 12,000 years ago. Pigs were next to be domesticated and then around 10,000 years ago, *Bos taurus*, or cattle, were domesticated. Ancient DNA studies show that as agriculture spread out from the Fertile Crescent to Europe, farmers brought *Bos taurus* with them, probably because the domestication of the local wild cattle in Europe was somewhat intimidating; aurochs (*Bos primigenius*) stood over 1.5 metres (5ft) tall at the shoulder and wielded horns nearly a metre (just shy of 3 feet) long. By 8000 BC, cultures across Europe and Asia were sowing a variety of seeds and herding numerous animals: in short, they were farming. It's unclear exactly when and where these first farmers decided to pull on the teats of these domesticated animals, but milk residues first start to show up in 9,000-year-old pottery recovered from archaeological digs in the Near East. And we know this residue is milk thanks to some very sophisticated science.

Richard Evershed FRS, Professor of Biogeochemistry at the University of Bristol, with whom I was fortunate to co-author my first book, developed the methods that have enabled scientists to trace the origins of dairying in ancient cultures. Pottery from this period was unglazed and therefore very porous. Any liquids placed in the pottery soaked into the clay itself. Thousands of years later the organic residues left by these liquids, though degraded with time, can still tell us a great deal about what these pots contained.

In the late 1990s, Evershed and his collaborators analysed more than 1,000 potsherds recovered from archaeological sites across Europe and although it was easy to identify the organic residue as fat, analytical methods hadn't yet been developed to figure out whether it was body fat (adipose) or milk fat.

And now for a quick briefing (I promise) on fat. All fat has a 'signature', which can be used to help identify its source. This signature is based on the composition or profile of the different fatty acids that make up the fat. The main constituents of all fats, including vegetable oils, are triacylglycerols (also known as triglycerides), which are composed of three fatty acids connected by glycerol. Each fatty acid has a carboxylic acid at one end, bonded to a chain of carbon atoms with hydrogen attached. As food consumers, we hear about fatty acids quite a bit – usually in terms of the health benefits of polyunsaturated fatty acids, including essential fatty acids such as omega-3s and omega-6s. There are cautions about saturated fatty acids and full-blown warnings about trans fats. The difference between these fatty acid types is simply that saturated fatty acids have single bonds joining the chain of carbon atoms, unsaturated fatty acids contain one double bond somewhere in that chain and polyunsaturated fatty acids contain more than one double bond. The number of double bonds and the number of carbon atoms in the chain differentiates each fatty acid, and affects its properties and how it is used within the body. Therefore, every source of fat will have different properties depending on its fatty acid profile. Animal fats, for example, contain more saturated fatty acids, which makes them solid at room temperature. However, vegetable fats have a higher proportion of polyunsaturated fatty acids with multiple double bonds, which makes them liquid at room temperature.

Had the pottery Evershed and his team analysed been freshly knocked off the kitchen counter, they could have used this fatty acid profile to differentiate the fat as being from a milk source or from adipose. However, as it turns out, milk fat degrades relatively quickly and soon takes on a similar composition to adipose fat. Evershed and his colleagues had to find a different analytical approach.

Conveniently, the fatty acids in milk come from a slightly different source from the fatty acids in adipose fat and these differences can be exploited in order to tell them apart. In a lactating cow, for example, about 40 per cent of the milk fat comes directly from what she is eating – in this case, grasses. Bacteria in her rumen, the first stomach of a ruminant animal, break the fatty acids free from the glycerol and these free unsaturated (because they come from plants) fatty acids combine with hydrogen found in the acidic environment of the rumen. This converts them into saturated fatty acids and they are then diverted to the mammary glands, where they are incorporated into her milk.

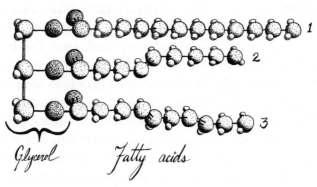

Figure 2.1 Triacylglycerol is made up of a glycerol, which binds three fatty acids. In this hypothetical triacylglycerol, fatty acid #1 is a saturated fatty acid with no double bonds, fatty acid #2 is an unsaturated fatty acid with a single double bond, and fatty acid #3 is a polyunsaturated fatty acid with more than one double bond.

The fatty acids in adipose fat, on the other hand, are mostly made from acetate, which is produced when bacteria in the gut ferment carbohydrates (mainly fibre). Compared with fat from the same plant, carbohydrates happen to contain a higher proportion of carbon atoms that have an extra neutron (known as the ^{13}C isotope). So in a very simplified way, when a lactating cow eats grass, fatty acids from the grass are diverted to her milk while carbohydrates, containing heavier carbon isotopes, are diverted for storage as body fat. Because of this, Evershed and his team hypothesised that adipose fat should have a higher ratio of this heavier carbon isotope than the milk fat. Evershed and his team tested this using modern cud-chewers (ruminants). It worked. They could easily distinguish the two types of fat based on the carbon stable isotopes of the fatty acids. They applied the method to the potsherds and confirmed that indeed the organic residues were consistent with the isotope signature of milk fat. It was the first conclusive evidence of dairying.

Our ancestors began milking their domesticated animals around 9,000 years ago and putting the milk in containers, but they couldn't actually drink it without getting an upset stomach. After the age of about eight, these Neolithic humans, like all mammals, stopped producing lactase, the enzyme needed to break down the milk sugar lactose. An inability to digest lactose leads to numerous uncomfortable stomach complaints, from cramps to diarrhoea, as anyone with lactose intolerance today can confirm. But these first farmers clearly recognised the nutritional benefits of milk as they watched babies (their own and those of their domesticated herds of animals) growing quickly on this liquid lunch. And they must have seen the benefits of a food product that didn't require killing the animal, otherwise why were they bothering to collect milk? So how did they get around the stomach cramps?

Archaeological evidence suggests that they were processing the milk into cheese and possibly yogurt in order to reduce the lactose content. Evershed and his colleagues have identified milk fat residues in pottery pierced with holes, which archaeologists presume was used as a type of strainer. And the logical reason for straining milk is to separate the curds from the whey.

There has been much speculation as to how Neolithic farmers stumbled upon the first curds. One theory is that milk was being stored or carried in sacks made from a ruminant's stomach when some chemistry magic inadvertently occurred. You can easily imagine the scenario. There is a drought and nearby freshwater sources run dry. A woman desperate to quench the thirst of her children milks her goats. She carries the milk in a skin sack (stomach), and as she walks home the sun warms the sack and enzymes still present in the skin sack curdle the milk within. It's a plausible story. In fact, it's the very story described in Greek myths. Except, instead of a desperate woman, it's the giant cyclops Polyphemus, the son of Poseidon, who discovers that the milk in his skin sack has curdled in the heat. By the time Odysseus and his crew discover the cyclops' cave in Homer's *Odyssey*, Polyphemus is clearly an accomplished cheesemaker. Odysseus discovers 'flat baskets laden with cheeses' and 'large and small pails, swimming with whey' – probably the beginnings of feta. Who knows what Polyphemus could have accomplished as a cheesemaker had Odysseus not poked his eye out!

My husband, a fellow biologist, has a slightly more gruesome theory, in that young animals would probably have been hunted (acknowledging that the old, young and sick make easier targets) or slaughtered from the domestic herd on occasion. As nothing would have been wasted, the stomach would have been opened or used, and inside the unweaned animal's gut would have been curdled milk. Someone decided to eat it and maybe realised it didn't have the same side effects as raw milk. It's an equally plausible story, although I can assure you that it would not have been me reaching into the emptied stomach contents of a slaughtered animal to pick out some curdled milk to pop into my mouth. But alas, this is my twenty-first-century perspective – where food waste is ubiquitous.

Perhaps the first curds were more simply a result of milk in a container getting warm and becoming acidified by the bacteria cultures within, and enzyme-induced curds didn't come until later. We will never know how the first cheese curd was discovered, but cheese-lovers today can be very grateful that our farming ancestors were experimental enough to give curdled milk a go.

The milk mutation

Cheese and other fermented products helped our farming ancestors to get around their lactose intolerance. By straining the curds, the majority of lactose is cast off in the whey, leaving the valuable protein (mainly casein) and milk fat in the curds. Fresh cheese curds would have provided a valuable source of nutrition. If they were further fermented or aged, they provided a method of storing this protein and fat for longer periods to provide a stable source of nutrition throughout the year. Cheese products probably became the currency of dairying economies in the Neolithic. More significantly, however, cheese possibly paved the path for changes in the human genome that persist today. And to understand how, we need to begin in the present and work backwards.

Approximately 35 per cent of adults around the world today are able to digest milk because they continue to produce lactase into adulthood – a condition known as lactase persistence. This varies considerably between different populations: 89–96 per cent of people in the British Isles and Scandinavia are lactase persistent, 62–86 per cent in central and western Europe, but only 15–54 per cent in eastern and southern Europe. Lactase persistence is very rare in east Asia, but there are hotspots of prevalence in west Asia, for example in Saudi Arabia, Iran and Pakistan. It is patchy in Africa, with a few areas with high prevalence (more than 90 per cent), particularly in eastern countries such as Ethiopia and Somalia and in north-western countries such as Mauritania and Mali. And only about 21 per cent of aboriginal North Americans can tolerate lactose.

In the last decade, a number of different mutations have been discovered in association with lactase persistence and it turns out that one mutation in particular, known as 13910*T, is associated with almost all lactase persistence in European populations. This mutation doesn't even occur in the lactase gene itself, but within a non-coding region of a neighbouring gene. The substitution of a single base pair in the sequence changes how the DNA is read during transcription. It's much like being given the instruction to transcribe the second sentence in a paragraph, but someone has removed all the punctuation. In the case of the lactase gene, this mutation causes it to be transcribed long after it would normally be turned off naturally with age. Researchers have analysed ancient DNA from Neolithic skeletons in Europe to look for the mutations associated with lactase persistence and not found them, which is why it is thought that they were lactose intolerant. This is supported by estimates of the age of these mutations, which indicate that the lactase persistence mutations we have identified today started to show up at about the same time as farmers started dairying.

So how do we go from nobody tolerating milk to 96 per cent of Brits capable of glugging it down over a mere 10,000 or so years? And, which came first − did people with lactase persistence start dairying or did dairying lead to lactase persistence? It's the chicken and egg question. Except it's not.

In terms of an evolutionary time scale, 10,000 years is lightning speed for a mutation to become so well established within a population. This suggests that there must have been a selective advantage in having this mutation and there can only really be such an advantage if the culture is already using milk. There are a number of theories as to what selective advantages the ability to drink milk would have provided. There is the calcium assimilation hypothesis, which notes that lactase persistence is more common in high-latitude regions of Europe. People living at these higher latitudes may have had less exposure to sunlight and therefore were unable to synthesise enough vitamin D. Vitamin D is essential for calcium absorption and, as it turns out, milk is an excellent source of both calcium and vitamin D. Yet this doesn't explain the lactase persistence hotspots of west Asia or Africa where sunlight wouldn't have been a limiting factor. In these countries, perhaps milk was an important source of uncontaminated fluids during periods of drought – something most northern European populations wouldn't have experienced. Different theories for different populations.

The fact is, of course, is that milk provides a lot of calories with essential protein and fat. It could have helped to see people of any region through seasons with low crop yields and to fill nutritional gaps between harvests. Most likely, all of these theories were relevant to varying degrees during different time periods and in different places. And although we may never be able to point to a single advantage, we can use some fancy models and maths to estimate how strong the selection pressures were on this mutation. As it turns out, the selection strength of lactase persistence comes out as among the highest estimated for any human gene in the last 30,000 years. It is thought that people with a lactase persistence mutation would have produced 19 per cent more offspring compared with someone without the mutation. So, while Marvel comics and blockbuster films imagine mutations involving telepathy, advanced cellular regeneration and skeletal claws, it is actually a mutation which led to milk tolerance that may have been the real mutant superpower in recent human evolution.

The selective pressure on lactase persistence suggests that dairying came first. The milk was processed into digestible products – cheese and yogurt – that formed the basis for a dairying culture. A few people within these first dairying communities of south-eastern Europe would have acquired lactase persistence through random mutation and been able to drink milk into adulthood without any ill effects. Natural selection would have acted on these few individuals who could access this nutritionally rich food more readily. There would have been a slow increase in the frequency of the mutation as these individuals were more successful (which in genetic terms translates to having more grandchildren than your lactose-intolerant peers). After this, the frequency rose more rapidly, co-evolving with the expansion of dairying across into central Europe. By the Middle Neolithic, dairying cultures were well established in Europe, as was the lactase persistence mutation. So you see, without cheese, these Neolithic cultures would have been challenged by bloating, gas and diarrhoea and would probably have given up on the whole dairying idea. Indeed, it was cheese that helped to shape the DNA of all of us lactose-tolerant descendants of these successful cheesemakers.

Diverse dairy: the Camembert explosion

The basic formula for cheese making hasn't changed in thousands of years, but there are factors at each stage that can be fiddled with in order to produce a unique product. There are also fundamental differences in the surrounding natural environment that ultimately shape the end product. This is known as *terroir* in the wine world, but it's equally applicable when it comes to traditional cheese.

Milk: This is the foundation. The source of the milk – goat, sheep, cow, buffalo, donkey, camel – will affect the final product. So will what those animals are eating – the quality and quantity of their forage. Cow milk yields a higher quantity of cheese than goat milk, but in some mountainous regions goats are the more appropriate animal to keep. So, here is where we see the first regional differences.

Starter cultures: Gentle heating of the milk encourages the growth of bacteria known as lactic acid bacteria (LAB), which acidify the milk by converting the milk sugar lactose into lactic acid. This acidification helps expel the whey from the curd, but it also makes the milk a lot less desirable for other bacteria. Before the days of pasteurisation and sterilisation, LAB would have been in abundance in the milk itself, in the previously used pots, in a ladle of yesterday's milk, and even on the cheesemaker's hands and the animal's udders. There are a number of different genera that can use lactose as an energy source, and the composition of these communities would have varied in different regions, potentially resulting in slight variations in the end product.

Coagulating the milk protein: The casein protein in milk is found in small aggregates known as micelles. Locked within these micelles are minerals such as calcium phosphate. The micelles prefer water to each other (they are hydrophilic), which keeps them suspended in the water component of the milk. When rennet is added to the milk, the active enzymes, chymosin (or rennin) and pepsin, cut the casein protein at a specific position, much like tiny molecular scissors. This causes the micelles to have a sudden aversion to water (they become hydrophobic). As a result, the casein micelles start to collect together, avoiding the water and forming curds. This can also happen without rennet, simply through the acidification of the milk, as anyone who has seen sour milk will know. Acidification causes the micelles to destabilise and aggregate, but it also increases the solubility of the minerals contained within these micelles. The minerals get released into the whey, which results in a more delicate curd and ultimately a more brittle cheese. Rennet creates a firmer curd and more of the whey can be expelled earlier in the process, retaining more of the minerals within the curd. So, there are even different variations in how the curd is formed. The Fulani people of northern Benin prepare a fresh cow's milk cheese known as *wagashi,* using the plant commonly known as Sodom apple (*Calotropis procera*). The milky sap of the fruit of this plant contains a complex mix of chemicals that make it impossible to get the sap off your hands and if ingested, it can be quite toxic. The Fulani, however, mash up the leaves to release vegetable rennet into the milk. Using this vegetable rennet creates a much harder, more gummy cheese curd, which means that *wagashi* cheese doesn't melt. The Romans also used plant rennet, mainly from fig and thistle, for their cheese making.

Capturing the curds: How the curds are treated is entirely dependent on how much whey the cheesemaker wants to extract from them. The curds for a soft cheese are often just captured in a mould and the whey is strained off through holes. This way a portion of the whey is retained in the curds. Prior to refrigeration, curds would have been dried, fermented or salted in order to preserve them. Tougher curds may be vigorously pressed to create a hard cheese or less vigorously for a semi-hard version. This is another step where tinkering can occur depending on the desired end product or the tools at hand.

Ripening the cheese: This is, without a doubt, where the majority of tinkering happens – and, much like ageing wine, both nature and crafters have a hand in this. Fresh cheeses aren't ripened, but even then there are variations in how the curds are treated. They may be cooked and stretched like a mozzarella, put in brine like a feta, or just stored in a pot like a ricotta or *queso blanco*. But it is among the ripened cheeses that true variation can happen. This is where further biochemical reactions are carried out through the growth of microorganisms in and on the cheese. The amount of water in the cheese, the temperature at which it is stored, salt levels and the pH of the cheese will all affect the growth of these microorganisms. Enzymes inherent in the milk, left over from the coagulation process and released as microorganisms die and break open, are critical in helping to break down the cheese components. The microorganisms and enzymes in the cheese break the protein down into smaller and smaller molecules and break the fatty acids free from the triglycerides. The amino acids and fatty acids contribute to the flavour and aroma of the cheese but they may also be converted through further biochemical processes into other flavour compounds. How the cheese is prepared for ripening, what it is inoculated with, where it is stored and for how long, will all affect the final product.

The first dairying cultures would have tinkered with these different steps, depending on their needs and the resources around them. We can gain some insight from the fragments of pottery these ancient cultures left behind and make suppositions based on what resources we know were to hand. But we can also look at how traditional cheeses are distributed around the world today to help us understand how and why different cheese varieties evolved. I acknowledge that these are broad, sweeping generalisations and that cheese connoisseurs will be able to point to exceptions in every case, but they are useful observations in describing how the process has developed.

Goat cheeses, not surprisingly, are more common to regions where cattle would have plunged to their deaths or starved to death. Despite cattle surpassing sheep and goats as the predominant herd animal in much of Europe during the Neolithic period, in dry and/or mountainous areas, such as much of the Mediterranean, goats persisted as the animal of choice. When you visit these areas, the reasons become quite obvious. I have personally scrambled around some of the countryside of Crete and it is, without a doubt, goat country. There are sheer cliffs and deep canyons, and the vegetation is sparse, dry and usually very spiky – on the whole, quite unpalatable. This is a land where only the resilient survive, and here I refer to both the goats and the

Cretans themselves. The kri-kri or Cretan ibex (*Capra aegagrus cretica*), which now roams the Cretan mountains in limited numbers, is thought to be the feral descendant of a domesticated goat brought to the island during the Minoan civilisation more than 5,000 years ago. It is domestic goats that still dominate the cliffs today and unsurprisingly, it is goat cheese (the Cretan version is known as *myzithra*) that dominates the traditional menu of Crete. This preference for goat cheese in mountainous regions or areas with poorer forage is a generalisation seen around the world, including *banon* from the French Alps, *rubing* from the mountainous Yunnan province of China, *tulum* from the mountains and plateaus of Turkey, and *picodon* from the dry hillsides of France's Rhône valley with its sparse vegetation.

Mountainous regions can also have shorter growing seasons, so many of the cheeses that evolved in the mountains may have been processed to create a longer-lasting product to help herders get through leaner times. *Tupi*, for example, is made from the milk of cows, goats or sheep grazing in the Pyrenees of Catalonia. It's named after the style of clay pot in which it is kept. Alcohol distilled from anise, known as *cazalla*, is added to the cheese curds and after a few days of stirring, the pot is covered and stored in a cool dry place for at least two months while the cheese ferments. The end product is quite pungent, but it doesn't have to be refrigerated and, most importantly, it lasts. *Chhurpi* or *durkha*, which is known as the hardest cheese in the world, originates in the Himalayas. It is made from yak milk and every ounce of moisture is squeezed out of the curds before they are smoked to make them hard as rock. This final product is usually called *chhurpupu* and the hard little blocks of cheese can be stored in a yak skin for up to 20 years. It can take hours to munch through a single slice, providing the chewer with some essential protein and fat as well as a set of formidable jaw muscles.

While some of the most aged cheeses evolved in places where growing seasons were short or unreliable, those ripened in large rounds could only have evolved in regions with stable, cooler temperatures – mountains or caves, for example. Warmer temperatures speed up the ageing process and prevent a large cheese from maturing evenly across the round, so before refrigeration, storing cheese in a large round in a warm place would have spelled disaster. Gruyère, aged in the Swiss Alps since at least the second century, is ripened in 35kg (77lb) rounds. The Franche-Comté region of France, which is bordered by the Jura mountains to the south and Vosges to the north, gives rise to Comté cheese that is aged in rounds weighing up to 50kg (110lb). And cheddar cheese has been aged in 28kg (62lb) rounds in caves for hundreds of years.

While salting curds is common practice, monks in Northern France and Belgium took this one step further and introduced the practice of repeatedly washing the cheese in a brine during the initial phases of ripening. This saline solution keeps the rind soft and prevents undesirable moulds that thrive in moister climates from growing, while promoting the growth of desirable bacteria, such as *Brevibacterium linens*. The bacteria ferment the cheese from the outside in, forming a rind on the cheese that protects it during this process. Cheeses produced in this way are known as washed-rind cheeses, and examples include Port Salut from the west coast of France and Pont-l'Évêque from the coast of Normandy. Washed-rind cheeses are quite renowned for the sticky reddish-orange skin formed by the bacteria, but they are perhaps better known for their smell. This is because *B. linens* also happens to be one of the species of bacteria that lives on human skin and is responsible for foot odours. Yes, that is correct – we encourage stinky feet-causing bacteria to grow on our food on purpose! The thing is, if you can get past the smell, you are usually rewarded with a pungent, earthy flavour that is unique and worth the nasal affront.

Fresh cheeses are most common in warmer countries where forage for domesticated ruminants is often more reliable. Rather than long preservation times, the main goal of processing was to reduce the lactose content as well as, perhaps, to make it more portable. Greek feta, Indian paneer, Egyptian *areesh* and Italian ricotta are all examples of fresh cheeses from warm climates.

And so, with a few tweaks of a basic recipe, a diversity of cheese products have evolved around the world. Trade and improving transport have helped to move these cheese products around, influencing cheese-making methods globally. Immigrants have brought their cheese-making skills to new lands and tweaked their processes to accommodate local resources, evolving new versions of old recipes. All of this contributed to a diverse global dairy landscape. And then the industrial revolution and war changed this landscape for ever.

The cheddar expansion

Cheddar is one of the world's most ubiquitous cheeses. In the UK alone, we consume 250,000 tonnes of the stuff annually. Americans produce approximately 1.4 million tonnes of cheddar each year. It is named after the village of Cheddar, which lies roughly 29km (18 miles) from my home office in Bristol. My family and I go there every time a new Canadian friend or relative comes to visit us. Even without the draw of cheese, this village would still be a destination for tourists as it sits nestled at the base of what is considered one of the finest limestone gorges in Britain. The gorge has been carved out over the last million years by boulders and gravel that were carried along by glacial meltwater, which flooded the landscape during the milder periods of glaciation. Today, the only rivers in the 4km (2.6-mile) gorge lie underground. It is now a well-travelled road that winds along the base of the gorge, where drivers attempt to navigate sharp turns while spotting feral goats that dot the limestone cliffs rising to 137 metres (450 feet) above.

It is within these limestone cliffs, in caves carved out by acidic rivers, that an ideal environment was found for

ripening cheddar cheese. At a constant temperature of
11°C, these caves provide a stable, cool environment in the
absence of refrigeration. Some cheddar is still being aged in
the caves as well as in the nearby caves of Wookey Hole
(you have to love British place names) and I get some every
time I go. It's the type of cheddar that is gritty, due to salt
crystals, and crumbles rather than slices. When you eat it,
it feels as though it's taking a layer of skin off the inside of
your mouth. It is delicious!

There's even an opportunity to pop into the Cheddar
Gorge Cheese Company's facility to watch them making
cheese. Although the room has flashy stainless-steel
equipment, it is still very much a hands-on process. Literally.
The cheesemakers still do most of the tasks by hand. Salt is
spread across the curds by hand. The consistency is tested
by hand so the cheesemaker can 'feel' how the cheese
process is progressing rather than relying on some automated
machine to tell them. They stack the curds by hand, which
is known as 'cheddaring', again to get a feel for the cheese.
The rounds (known as a truckle) are wrapped in muslin
and greased to prevent them from drying out. They are
then stored in their warehouse (only a few make it to the
caves) for nearly two years. This commercial-scale
production with processes based in tradition is a common
theme in the history of cheese production in Britain. So,
how is it that the humble cheddar, from the English county
of Somerset, became one of the world's most popular
cheeses?

In the early eighteenth century, cheese making was still
largely a farmhouse affair around the world: excess milk
was turned into cheese by local producers and sold in the
local area. There were, of course, some exceptions. For
example, cheese made from the milk of animals grazing
the Cheshire Plains in England was highly desirable and
demand forced Cheshire cheese products into commercial-
scale production. Dairy farmers pooled their milk and the
cheese was made in a few specialised cheese dairies rather

than each of the farmers making their own product. It was estimated that by 1729, 5,954 tonnes of Cheshire cheese was being shipped into London annually. Yet, despite such levels of production, traditional cheese-making practices were used and milkmaids and farmers' wives undoubtedly got very little sleep.

As the nineteenth century approached, the industrial revolution began attracting people into cities to work in factories. The milk followed them there. It was transported by steam train into these urban centres from the surrounding countryside to supply the masses. There was now an opportunity for urban-based dairies to use any excess milk to make cheese where the people were. They found that by pooling the milk from a much broader geographical base, they ended up with a more consistent product, which retailers and wholesalers liked.

Entrepreneurial farmers raced to increase production. In Somerset, dairy farmers like Joseph Harding were experimenting with cheddar production to make it more efficient in order to serve the evolving London market. These farmers tinkered with all the variables, from acidity to temperature, and they shared their findings with anyone who would listen. They not only trained fellow cheesemakers, they also published the results of their experiments in specialist journals. Harding and his contemporaries developed a steadfast system of making cheddar and it caught on with other cheesemakers beyond Somerset's borders. It eventually reached as far north as Scotland. It was just as Harding had hoped. He wanted to strengthen British cheese production to compete with imports from the Netherlands and Ireland and now he had a solid system for producing this hardy cheese.

But it was indeed a little too dependable. The cheddar process was exported around the world, particularly to Australia, New Zealand and North America. Cheddar can be ripened in huge rounds that are relatively resistant to extreme temperature and humidity fluctuations. It can also be consumed in a much milder form if ripened for less time. It was an ideal cheese for a variety of climates with a foolproof process for commercial production. And while cheese factories hadn't really caught on in Britain, by 1864 there were over 200 cheese factories operating in New York state alone. Only a decade later, there were 900 cheese factories operating in New York; they were supplying 90 per cent of the state's cheese needs and exporting over 46,000 tonnes of cheese back to the UK. Cheddar was no longer from Cheddar.

Recognising that they were being outdone, the first British cheese factory opened in Derbyshire in 1870, producing Derby cheese, a mild cheese akin to cheddar. Yet factory cheese failed to thrive in the UK, for many reasons, including a fierce loyalty to farmhouse-produced cheese. Despite this loyalty, however, retailers wanted consistent mass-market products, which they could sell at a great price. By the early twentieth century, British consumers were making sandwiches with uniform factory cheddar from Australia, New Zealand, Canada and the US. Less than a quarter of the cheese consumed by Brits was produced in Britain itself. Consumers with healthy purses and a pride in all things British protested against the shift to foreign factory cheese and touted the advantages of locally produced specialised cheeses – a mantra that continues to this day for any number of products. But these consumers were in the minority. At the start of the twentieth century, there were over 3,500 farmhouse cheesemakers registered in the UK, and by the 1930s this had dropped to just over 1,300. By the end of the Second World War there were barely 100.

War prevented the manufacturing of non-essential dairy products. The British government allowed only a standardised government recipe cheese, a mild white cheddar, to be produced. It fitted well with rationing schemes, but quality fell victim to quantity. The US government did the same. These controls weren't lifted in the UK until 1954. There has been a steady rise in the number of artisanal cheese producers since then, but even as we entered the twenty-first century, a mere 2 per cent of British cheese was produced by small specialist cheesemakers. However, at least the trend is headed in the right direction.

So, between wartime controls and the enthusiastic dissemination of tried and true processing methods, cheddar travelled the world and became nearly synonymous with the word cheese. Yet, as anyone who has tasted a good vintage cheddar knows, not all cheddars are equal.

To distinguish their product from others, producers in south-west England, who use traditionally based methods to produce a superior cheddar, sought Protected Designation of Origin (PDO). In Britain there are currently 12 protected cheese names, including Single Gloucester, Blue Stilton, White Stilton and Staffordshire cheese. But, because cheddar is such a widely used term, they could never be granted PDO protection for 'cheddar'. In 2007, however, they did get EU PDO protection of 'West Country Farmhouse Cheddar'. And, as with any PDO designation, there are strict rules and auditing processes in place to ensure the cheese is produced in a particular way. The milk must come from local herds reared and milked in one of the four counties of Somerset, Dorset, Devon or Cornwall; it doesn't contain any colouring, flavouring or preservatives; it's made using traditional methods, including hand cheddaring, on farms in those same four counties; it is matured on those farms for at least nine months; and it contains at least 48 per cent fat and at most 39 per cent water.

Beyond this designation, however, there is all manner of cheese out there that is called cheddar – from an unnatural orange cheddar sold in Canada and the US to pre-shredded stuff that's more vegetable oil and cellulose than dairy.

Most mass-market cheeses are a far cry from the crumbly tangy delight I buy whenever I'm in Cheddar. Many people within the industry now distinguish these as modern-style cheddars versus the traditional cheddars. The former is a product shaped by production costs and demand for marketing consistency. The latter has evolved over hundreds if not thousands of years of craft and a desire to produce a unique-tasting product that consumers want to eat. And it is this distinction that leads us to the final part of this chapter. At some point, cheese became no longer about providing a unique experience for the consumer, but rather an exercise in producing mass quantities of a safe and consistent (and quite frankly bland) product. Now, manufacturers are trying to achieve both, but is that possible?

The cheddar approximation

It took humans approximately 8,700 years to go from salvaging some curdled milk to crafting hundreds of different types of cheese around the world. It took us less than 200 years to then turn a good portion of that cheese into a monotonous mass-produced proximity of the real thing. And the fundamental shift that thrust us on this giant leap backwards is that commercial manufacturers began to question how they could make a more consistent product with less fat, in less time and for less money. This is the same story for many mass-produced foods and the shift rarely happens overnight. People tinkered little by little and these small changes over time ultimately led us to the shrink-wrapped bricks that adorn cheese aisles everywhere.

We can look as far back as the seventeenth century to see some money-saving efforts that helped to shape modern cheese. English dairy farmers found that they could make more money by skimming some of the cream from their milk to sell as it is, or as butter. However, the carotenoid pigments in the cow's diet that give cream its colour are fat soluble. So, when some of the cream was removed, so was the rich yellow colour that customers associated with a quality full-fat cheese. Instead of adding the fat back in, the farmers began to fake it by adding marigold or carrot juice. By adding the colour separately, the cheese became consistently coloured through all seasons, whereas normally it would change as the cow's diet changed. Some cheesemakers began using the vegetable dye annatto, which is used to make margarine and custard a richer yellow, to colour their cheese to hues that were clearly not from milk alone. Consumers were trained to consider richer coloured cheeses as superior products and it also helped to set certain styles of cheese, such as Red Leicester and Double Gloucester, apart.

The practice of adding annatto to cheese was brought to North America and by the early 1900s, cheesemakers in Canada and the US were adding annatto to their version of cheddar. Head into a Canadian supermarket today and you will be blinded by pumpkin-coloured blocks of cheddar. There's even a marbled version where orange-dyed curds are mixed with white curds to create a kaleidoscopic sandwich-filler.

Coloured cheddar has its limits, however. After about two years, crystals form in ripening cheddar. These are delightful crunchy bites of calcium lactate that form through the break-down of lactose and are a general indicator of a well-aged cheese. While these crystals aren't noticeable in a naturally coloured cheese, they become quite obvious when it's bright orange. This is why orange cheddar is never aged for more than two years, and usually far less.

In Illinois, at the turn of the twentieth century, an entrepreneurial man by the name of James Lewis Kraft was about to take orange cheese to a whole new level. In 1916, Kraft was awarded a patent for the 'Process of sterilizing cheese and an improved product produced by such process'. In his forties by then, the Canadian-born Kraft had worked his way up from a grocery store clerk in Fort Erie, Ontario, to co-founder and president of J.L. Kraft & Bros., cheese manufacturers in Illinois. Kraft used extreme heat to convert cheddar into a product that could endure an apocalypse 'without substantially impairing the taste' of the cheese. Anyone who has had a Kraft slice is welcome to judge whether Kraft achieved his latter objective. Kraft's challenge was to heat that cheddar up high enough to kill any living thing within it without melting it to the point where the fat separates irreversibly from the protein (picture the pools of fat that accumulate in the hollows of stringy melted cheese on a hot pizza). Kraft overcame the challenge through agitation and stirring. In other words, if he whisked that cheese constantly, he could bring it beyond its melting point to 80°C (175°F) for 10–15 minutes in order to properly sterilise it without it separating. Then it was a matter of cooling it and sealing it. The patent was awarded for the process and the product, which is described as 'a hermetically sealed completely sterilized package of non-liquid homogeneous cheese of the Cheddar genus'. The invention helped launch the Kraft cheese empire and it is still the cheese of choice for most fast-food burgers.

Kraft's patent was essentially a modification of a process developed in Switzerland by Walter Gerber and Fritz Stettler (of Gerber and Co.) in 1911. Gerber and Stettler shredded and heated Emmentaler cheese along with sodium citrate, an emulsifying salt, to produce a long-lasting heat-tolerant processed cheese.

It was Kraft's brother and business partner, Norman, who developed the technology to make the non-liquid homogeneous cheese even more convenient for consumers. In 1935, Norman began experimenting with the liquefied cheese to see how he could get it to consumers pre-sliced. He poured it out onto a cold stainless-steel table and flattened it with an ice-cold rolling pin, and found that he could easily cut it. Norman and the Kraft engineers set to work building a machine that automated the process and in 1944 they were awarded a patent for their invention. The machine squeezed the liquefied cheese out between two chilled rollers that pressed it into a thin sheet on the surface of the big lower roller. At this point, the sheet of cheese was cool enough to not spread any further than desired, yet flexible enough that it didn't permanently take on the curvature of the roller. A conveyor belt carried the cheese sheet away as it came off the bottom roller. Pizza cutter-like devices then cut the sheet into strips of a designated width. As the strips moved along the conveyor belt, a worker manually ran a cutting device across the strips to produce rectangles of a desired size. Other workers then pulled the slices of processed cheese off the conveyor belt and stacked them in a rack. The patent goes on to describe in mind-boggling detail (as patents tend to do) the angles of these racks and the precise method of trimming the cheese slices using trimming guides. I'll spare you these details. However, the patent does say that all the trimmings can be returned back into the cooker to be re-melted and start the cheese-slice journey all over again. Once trimmed, eight slices of precise thickness were wrapped up together in a package. It wasn't until the 1960s that Kraft developed the technology to wrap each of the slices individually in soft plastic – the height of convenience. And because the cheese was cooled quickly on the rollers, it took on a glossy, smooth finish. This finish helped prevent the cheese from drying out and getting that crusty rind that natural cheeses get. This shine also apparently gave it a 'quality of attractiveness', which no doubt confused harried mothers everywhere about whether they had actually removed the plastic wrapping or not.

Processed cheese singles are still alive and well today, providing the filling for grilled cheese sandwiches and toppings for cheeseburgers around the world. For some children, this could very well be the only cheese that they ever really know. That, or any number of lunchbox fillers that contain cheese dips, strings, twists and sticks, which are all descendants of processed cheese technology.

My survey of patents and scientific literature from the early twentieth century leads me to believe that efficiency was the major driver of cheese technology at the time – finding ways to reduce spoilage and make processes more economical. Patents for various machines that reduced the human contact time in the cheese-manufacturing process were abundant in the 1920s, 30s, 40s and 50s. This included better cheese cutters, better presses, and machines that put matting and moulding of curds into the hands of machines rather than cheesemakers.

As well as automation, there was a drive to shorten the ripening time required to produce a flavour that was acceptable to customers. Less time means manufacturers save on storage costs, but a stronger flavour means they can charge more for their product. All cheddars, supermarket brand or otherwise, have some sort of strength indicator as part of the label. In Australia, the cheddars are labelled as mild, matured, tasty or extra tasty. Here in the UK, they are labelled as mild, mature, extra mature or vintage and in Canada and the US, it's mild, sharp, extra sharp and seriously sharp. These designations are based on the flavour and texture of the cheese as determined by graders, rather than a specific period of time of ripening. The cheese will be monitored as it ripens and the graders will look at and smell the cheese and take small samples to determine how it is ripening. If it has all the indications that it is ripening well, it will probably be allowed to continue to do so. If, however, there are indicators that it is not following the predicted ripening pathway exactly, they will decide when to sell it, deciding on a point before it develops unfavourable flavours and when it will have the most value. A two-year-old cheese may be tested as many as six times during its ripening. Shortening the time to develop a decent flavoured cheese is the holy grail of cheese manufacturing.

In the early 1900s, it was well known that adding enzymes to the milk – beyond rennet – could help to shorten the cheese ripening time. This is, after all, what these specialised proteins do: they catalyse biological reactions. The enzymes in rennet cut the casein up into peptides. The bacteria cultures in the cheese then use enzymes to break these casein-derived peptides up even smaller, eventually producing amino acids (the building blocks for proteins) and other flavour compounds. A number of patents were published that modified the cheese-making process through the addition of enzymes that were often derived from macerated animal tissues that would have otherwise gone to waste. One patent from 1957 describes adding 'finely divided beef, hog or sheep kidney to the milk' – roughly 1lb of finely divided kidney to 1000lb of milk. But this was rather unconventional; usually the enzymes were isolated from the tissues long before they were introduced into the milk. However, adding enzymes is tricky as too much enzyme causes bitterness in the end product and changes the texture of the cheese because the rapid breakdown of peptides causes the protein matrix to collapse.

In 1949, on the heels of the Second World War, pasteurisation laws were passed in the US, which required that cheese should either be made with pasteurised milk or be aged for 60 or more days, ruling out unpasteurised soft cheeses. The results were safe, but admittedly a little boring. Using pasteurised milk to make cheese reduced some of the uncertainties and improved the uniformity of the cheese products, which the people who sold cheese liked. But it also led to slower ripening and a loss of flavour characteristics, which the industry was less fond of. Luckily, consumers had already been primed for this latter side effect with the mild cheese that had been rationed during the war – a cheese that was often referred to as 'mousetrap' in the UK.

By the 1950s, cheese in a jar had made it to the market. Cheese Whiz was America's convenient version of Welsh rarebit (a fancy cheese sauce on toast). Instead of cooking up a sauce of cheese, butter, mustard, flour and other ingredients, one only needed to spread some Cheese Whiz on a slice of warm toast. Cheese Whiz was never marketed in Europe. Instead, Brits had the Dairylea cheese triangle, which was introduced in 1950. Not exactly a rarebit replacement, but a processed cheese that could be spread on toast, and more often (according to my cousin Carol), it was simply unwrapped and popped directly into the mouths of children. The cheese triangle has been credited as being the UK's first fun food for children.

Other cheesy shortcuts were also being explored. A patent was awarded for imparting cheese flavour into baked goods without the use of cheese at all. Up until this point, very strong mature cheeses were needed to provide sufficient cheese flavour to baked goods to be even noticeable. As an example, it required 50lb of mature cheese to every 200lb of flour for a cheese cracker dough. This is very costly, and one might question why you wouldn't just put a slice of this good cheese on top of a plain cracker. But cracker producers wanted a savoury snack that stood on its own. In 1951, a patent was awarded for the process of adding the amino acid leucine to baked goods. Instead of 50lb of mature cheese, less than 1lb of leucine was required to give a cheese flavour. The only problem was that after two days the cheese flavour vanished. However, the inventors found that by adding a cheaper mild cheese along with the leucine, they not only got a more intense cheese flavour, it lasted the lifetime of the cracker. This was the beginning of cheese becoming an affordable ingredient.

The 1950s was also when modern gas chromatography was developed. This analytical tool enabled scientists to separate and identify different flavour compounds in foods, including cheese. They analysed the volatile compounds that wafted off the cheese at different stages of ripening in order to better understand the chemical processes that were happening. They analysed the amino acids found in the cheese to see which ones were responsible for imparting different flavours. Glutamic acid, for example, gives an umami (savoury) characteristic, while glycine is sweet, tryptophan is bitter and cysteine has a thoroughly unpleasant sulphurous flavour. Ketones, alcohols, aldehydes and esters that were significant to either aroma or taste were being identified as well. This gave food scientists a target for tweaking cheese-making processes in order to enhance the development of desirable compounds and restore some of the flavours that had been lost through pasteurisation. But sometimes that is easier said than done.

The 1960s and 1970s were an era of entertaining. Advertisements showed rows upon rows of carefully placed crackers with various toppings glued on with Cheese Whiz. Snack Mate (a processed cheese in a can) was advertised as the 'cheese that goes anywhere' − it could be sprayed out of a can onto hotdogs, macaroni, celery or even to top off a French onion soup. The marketing imagery centred on friends having a good time gathered around cheese plates. And if you didn't have friends or a significant other, then cheese might actually change that; an Irish campaign launched in 1970 called cheese 'manfood', known to make men feel good. The inference was that women could use Irish cheese to seduce any man!

For supermarkets, competitively priced cheese could be used to lure customers in through the doors, so they wanted a reliable product with better profit margins. Food scientists continued to look for ways to speed up the process, which would help profits but also help use up the abundance of unwanted milk fat that seemed to be accumulating. Health concerns over saturated fat that started in the 1950s meant fewer people were drinking milk in the 1960s, and if they were, they were choosing reduced-fat milk. Governments were buying up milk to prevent dairy farmers from going under. Europe was stockpiling 'butter mountains' while the US was hoarding cheese (of the processed variety so it would last longer) in the cool caves of Missouri. It was a time of rapid manufacture.

Commercial starter-culture companies developed methods to freeze-dry liquid cultures of bacteria so they could be added directly into bulk tanks to increase the bacterial cultures. More bacteria speed up the process, but they also produce more lactic acid, which is undesirable. So they acid-shocked or heat-shocked the LAB cultures, which slowed the little critters down enough to stop them producing lactic acid, while they still contributed desirable enzymes to the mix. These additional attenuated cultures helped speed up ripening, achieving fully aged cheese flavours in 10 weeks rather than 10 months according to some patent descriptions. In 1975, Kraft was awarded a patent for adding cyclic adenosine monophosphate (cAMP). This is a messenger molecule important in many biological pathways, including the regulation of a section of DNA in bacteria that is critical for the metabolism of lactose. The patent claimed that this made flavour develop in cheese two to five times faster than normal.

The industry was also making use of the new and improved understanding of flavour compounds; over 180 flavour compounds had been identified for cheddar alone. Patents were being filed for different combinations of chemicals that produced specific cheese flavours. For example, the right mixture of 2-heptanone, 2-nonanone, 2-heptanol, 2-nonanol, phenol, butyric acid, 1-octen-3-ol (also known as mushroom alcohol) and methyl cinnamate could be used to produce a Camembert flavour or a blue cheese flavour, simply by playing with the proportions of those last two ingredients. This list looks terrifying, but these are all organic compounds found in any number of foods that impart flavour; 2-nonanone is also found in bananas, ginger, Brazil nuts, corn and asparagus, while 1-octen-3-ol is a compound produced by plants and fungi and is found in our own breath and sweat. The names can be a bit off-putting, which is probably why they don't allow chemists to write food labels.

In the 1980s, concerns over saturated fat continued. Though, as it turned out, we probably should have been more concerned about the amount of hairspray we were using on a daily basis. But I digress. Low-fat cheese had been looked at back in the 1940s, but the major motivation at that time was to make more cheese using less fat. The result was a rather rubbery bitter cheese that didn't melt or grate well. In the 1980s, efforts were concentrated on improving these defects, which seem inevitable when you take the best thing about cheese ... out of the cheese. As well as the enzyme- and culture-tweaking that generations before them had used, food scientists added LAB cultures that produced carbohydrates that specifically bind water, which helped increase the moisture content of the cheese. They played with temperature and pH to try and control the development of bitterness and maintain moisture. Fat replacers were added. These included reduced-calorie ingredients made from whey, milk or egg protein, or even plant carbohydrates, which mimic some of the characteristics fat imparts to cheese, including viscosity and mouth feel. Cellulose was added to try to fill the fat void. Gums, carrageenan, gelatin and other substances that give low-fat yogurt that jelly-like texture were also experimented with to try and improve low-fat cheese. Personally, I am of the opinion that full-fat in moderation is the way to go.

While some people were trying to figure out how to replace the fat in cheese, others were trying to figure out how to replace the protein. The push for cheese analogues (fancy word for imitation cheese) was driven by several factors. It provided a cheese-like product in places where dairy is limited, or for people who wanted to avoid dairy products. But it also provided a cheaper cheese alternative that could be used as an ingredient and that could be engineered to have specific properties, such as melted stringiness for the top of a pizza, and fortified to provide nutritional benefits. And the processes and equipment for manufacturing analogues was extremely similar to those for making processed cheese.

A survey in the 1980s that compared cheese analogues to natural cheeses found that people were more willing to accept analogues in certain foods. Imitation cheese on pizza was absolutely acceptable, but it had to be the real thing for macaroni cheese. A slice of the fake stuff on top of a burger was fine, but it wasn't acceptable on a cracker. Imitation cheese in a sandwich was fine, but if that sandwich was grilled it needed to be real cheese.

The scientists went about finding a way to replace the milk proteins with vegetable proteins. The problem was, they didn't have the same properties. While casein was quite resistant to heating and drying and other processing, vegetable proteins were not. The vegetable proteins were a good nutritional substitution, but functionally they didn't work. The early analogues used some isolated milk proteins to maintain function, and hydrogenated vegetable oils as the fat source. This improved how the analogues melted, but they were still not good enough to be commercially successful.

Food scientists were also getting more sophisticated with their enzyme cocktails in the 80s, selecting specific enzymes that could help produce the precursor molecules that microorganisms then convert into flavour compounds. Essentially, scientists can tweak the system to exaggerate aspects of the flavour profile and they can tailor it to specific needs. Cheese curds are pasteurised to stop biological reactions and then blended with closely guarded concoctions of enzymes and a starter culture. The mixture is incubated for a few days before it is treated with heat to inactivate the enzymatic processes. The result is a paste that can be sold as is, or dried and sold as a powder to other cheese manufacturers. It's known as enzyme-modified cheese (EMC) and it can have between 5 times and 25 times the flavour of a natural cheese, though some have even claimed they can achieve 50 times the flavour. EMCs are used as an ingredient in any number of different foods, including soups, sauces, dips, crackers and snack foods. They can even be mixed with cheese powder, which is spray-dried processed cheese such as that used in packaged macaroni cheese, to boost the flavour. There is no need to wait for EMC to ripen, it's cheaper than natural cheese as you need less to create a similar amount of flavour, and it is lower in calories than a natural cheese. It's also cheaper than synthesising the flavour compounds in test tubes. More often than not, EMCs are used in addition to natural cheeses. A cheese filling for ravioli, for example, might contain 51 per cent (by weight) natural cheeses and 1.2 per cent EMCs, with the rest of the weight being mostly water and cream. It is the ultimate ingredient cheese.

With the 1990s came a fundamental shift in thinking within the industry. According to Barry Law, an R&D consultant to the dairy industry based in Australia and author of a whole lot of cheese-related books and papers, cheese manufacturers changed their attitude from 'you'll eat what we're willing to make' to 'we'll make something you want to eat'. Supermarkets wanted their own-brand products to stand out with unique flavours. The commercial cheese industry was emerging out of the safe and boring era into safe and moderately unique.

How they did this seems to be largely a matter of tinkering with the same things cheesemakers have been playing with for decades: enzymes and microbes. They knew far more about the microbes involved in the cheese-making process, which meant cheesemakers could fine-tune things like ripening temperature and moisture in order to maximise microbial activity without risking spoilage; instead of ripening cheese at 5°C, temperatures could be increased to 12°C or even 15°C for some cheeses. This decreased the maturation time by 60 to 75 per cent and reduced refrigeration costs at the same time. The relationship between the microflora and cheese flavour chemistry was also now better understood and so more cultures that could enhance specific flavour profiles without over-acidifying the cheese became commercially available.

The significant advancements in DNA-based methods in the 1990s and on into the twenty-first century has enabled cheesemakers to take fine-tuning to a whole new level. Not only is it now possible to choose different strains of microorganisms that seem to yield preferable cheese qualities, but these critters can now be manipulated to form 'super starter cultures'. Some bacteria, for example, carry mutations that cause them to produce more enzymes than non-mutants, making them better suited for cheese making. With DNA-based methods these mutants can be identified and cultured for commercial purposes. Genetic engineering can also be used to enhance desirable qualities in the bacteria. This usually involves introducing more copies of certain genes so that the bacteria produces more of a desirable enzyme or changes how rapidly the culture grows. This genetic manipulation is enhancing qualities that already existed in the bacteria, but people both within and outside of the industry get jumpy around genetically modified organisms (GMOs) and their associated stigma, whether justified or not.

Non-dairy cheese substitutes have finally come into their own with more demand for vegan products. There have been improvements in the products themselves, but there is also simply more customer acceptance that soy cheese behaves differently from a mozzarella on top of pizza or that cashew cheese is never going to taste like a vintage cheddar. But a vegan-friendly dairy-free cheddar style 'cheese' isn't pretending to be anything but an alternative. Cheese analogues – which by definition look like cheese – are used as ingredients in other products and it isn't always clear to the consumer that they aren't cheese. Pizza cheese is the classic example that has hit headlines every few years since the 1980s. If you see cheese on the top of a frozen pizza from the supermarket or melted on that take-away you ordered, you assume it is indeed cheese. And yet, a look at the ingredient list might reveal that it's listed as 'pizza topping', or, as numerous investigative reports have revealed, misleadingly listed as 'cheese'. This is the unfortunate drawback of creating such a believable substitute.

As always, however, the quest for ever-faster ripening has continued. A method for ripening cheese under high pressure was patented in 1993 by the Fuji Oil Company. The inventors claimed that by putting a cheese under enough pressure to speed up ripening without imploding all the microbes, they could achieve the equivalent of a six-month ripened cheddar in just three days. They just had to add a whole lot more microbes to begin with because bacteria can't proliferate under pressure, apparently. Other research groups used similar methods, however, and did not get such striking results. Fuji Oil Company's patent is active, but it's unclear whether it has amounted to much as despite a great deal of academic hype about high-pressure processing of cheese, all of the sources I have spoken to have said that it isn't commercially viable because equipment costs are too high. In 2004, however, a Canadian cheese company based in Québec got creative. The company dropped 800kg (1,763lb) of cheese to about 50m (164ft) depth in Saguenay fjord near the mouth of Rivière Saguenay north of Québec. The idea was that pressure and a somewhat unusual environment might develop a rather unique cheese. The problem was, they couldn't find it again. Despite professional divers and high-tech tracking equipment the cheese was never found. The company had to call off the search and write off CA$50,000 (£31,000) worth of cheese in the end.

Highly unique products seem to be the latest trend in cheese making, from sinking cheese in a river to ripening cheddar 152m (500ft) below ground in the old slate mines of Wales, much to the delight of experimental cheese consumers everywhere. There has also been a steady rise in the number of artisanal cheesemakers globally, which have been greatly supported by 'buy local' movements. Cheese is regaining its diversity.

The nutritional implication of this is mainly that we are eating an ever-increasing amount of cheese. On a global scale, cheese consumption has been on the rise by about 2 per cent each year. This is mainly because of its increasing popularity as an ingredient over the last 30 years. Cheese is in a good portion of the meals on any restaurant menu and it is the foundation of most menu items in popular Tex-Mex chains and pizza joints. These days, cheese (natural, processed, EMC and analogue) is baked into crackers and bread, mixed into sauces and stuffed into pasta; it coats popcorn, crisps and other snacks, and it's in any number of lunchbox-ready formats for kids.

This is all a long way from curdled milk in a skin sack. We have clearly gone far beyond just removing lactose and storing milk protein and fat, but have we gone too far?

There are stories of factories that have milk going in at one end and plastic-wrapped bricks of cheese coming out at the other. And this is true in some cases. Enzymes and microbes are used to help speed natural processes along and we do this in wine making, baking, production of juices and baby foods. The reality is that this quickly manufactured product is unlikely to ever achieve the depth of flavour that a truly aged cheese will, but that's OK because these mild cheeses have a role too. Nestled among the stinky speciality cheeses in our cheese compartment, there is always a block of mild or mature cheddar. This is what tops off our homemade nachos or tacos and fills my son's sandwiches as he has yet to appreciate the more pungent varieties.

As for convenience cheese, the stuff that comes pre-grated or individually wrapped or in various fun formats for kids doesn't work for me. Cheese tastes better freshly grated and it is the task that my son most loves when helping to prepare dinner as he sneaks a few pinches as his fee for helping. Pre-grated cheese also needs to be coated in cellulose, potato starch or some other added ingredient so that it doesn't all bind into a big ball once packaged; it's not harmful, but who needs it? The kids' cheese snacks are always over-packaged, individually wrapped in often non-recyclable materials. I'm not willing to sacrifice the environment or buy processed cheese products for the sake of a little convenience.

Imitation cheeses and cheese analogues provide an important non-dairy alternative for those who want them.

And if the availability of milk changes in the future, we have developed the technology to create cracker toppings and sandwich fillers for those who crave them. My concern with these products, however, lies in the potential for deception and fraud. They cost far less to produce and even if they are used to bulk out real cheese in a ready-made lasagne or frozen pizza, they help to substantially lower the production costs. Most countries have reconsidered their labelling guidance as a result. In the UK, for example, these products 'may not be labelled, advertised or presented using protected terms reserved for milk and milk products'. There shouldn't even be an indirect suggestion that these products are connected to dairy, which some argue includes having them beside dairy products in the supermarket. It most certainly means they should not be called cheese. Labelling must be clear: it should say cheese analogue or imitation cheese in the ingredient list, but if it is broken down into its constituent parts (water, vegetable oil, milk protein, starch, emulsifying salts, colour, flavouring) does the consumer stand any chance of recognising these as being a cheese analogue? From an animal welfare perspective and potentially an environmental perspective there may be huge advantages to using analogues, but the point is that people need to know and make the choice for themselves.

Ingredient cheese, particularly powders and enzyme-modified cheeses, provide a powerful cheese-flavoured punch with less cheese, less fat and therefore fewer calories. But fat is important: it's what causes the release of the hormone leptin in the body and tells the brain we're full. We think nothing of munching our way through a bag of Doritos, but a small chunk of cheese is enough to satiate. The world of flavours seems to be a grey area in my opinion. Whether they are from artificial or natural sources, it feels like the culinary equivalent of performance-enhancing drugs. No matter how hard you worked at a recipe, you simply could never achieve that intensity of flavour without them.

And then there is processed cheese: a cheese product originally designed to increase the shelf life of cheese, yet the amount of cheese in some of these products has slowly dwindled through the decades. Nutritionally, processed cheese sits within the range of other cheeses. In terms of macronutrients, the protein content of processed cheese is equivalent to a Camembert or Danish blue. It has the same fat content as fromage frais and significantly less than a cheddar or Stilton. It is lower in cholesterol than Parmesan or Gruyère and has the caloric equivalent of Edam. Processed cheese also sits in the middle of the road in terms of vitamins, containing a little less vitamin D than cheddar but more than mozzarella. Most processed cheeses contain emulsifying salts to prevent the fat and protein from separating during heating and so it is not surprising, perhaps, that it is relatively high in sodium: 100g can contain about 1320mg sodium, about two-thirds of the average recommended daily intake. Roquefort and feta both have more sodium per 100g than this (1670mg and 1440mg, respectively), although due to their richness, one might have a harder time eating as much of these cheeses in one sitting as you could a processed cheese. Processed cheese has more calcium than Brie or mozzarella, but not as much as Parmesan. And so there isn't a nutritional argument against processed cheese to speak of. Personally, I'm not a fan of the taste or texture (or lack thereof) of processed cheese. However, its exceptionally long shelf life does mean it has helped to fill a niche, it has been included in army rations and it has helped to store America's dairy surplus. And who knows, perhaps if things really go sideways here on Earth, processed cheese will be the only thing left standing. It will give the cockroaches something to eat – so long as they can figure out how to get that plastic wrapping off, of course!

CHAPTER THREE

Breaking Bread

My mother often reminds me of a story from my childhood where she had me sitting in the shopping cart as she went round the supermarket. She had put a baguette into the cart with me. This was a true treat – white bread – because as you will recall from my introduction, we generally had heavy homemade whole-meal at home. By the time my mother got to the checkout the baguette was significantly lighter. I had discreetly poked a hole in one end and mined out all the soft white insides, reaching in as far as my toddler arm could get me. I was too young to recall the incident and can therefore neither defend myself nor confirm the accusation. However, as I have been tempted on more than one occasion to do this as an adult, it is very likely true.

I feel a genuine sadness for people who have to avoid gluten. I tried once to see if I felt better without gluten; I lasted a day and went to sleep with bread-filled dreams. I had toast the next morning. In fact, as I write these words, I am privileged enough to be sitting in the south of France dining daily on croissants and baguettes. I am in my own personal food heaven. I have even managed to pass my love of bread on to my son who yesterday said to me, 'Mum, you know how carnivores are predators? I think I'm a breadator!'

If I ate any old pre-sliced stuff from the supermarket, I might feel differently, but it's almost always homemade at my house - a lovely light seeded wholemeal, or, lately, sourdough. Once a week the kitchen is filled with the smells of baking bread, more frequently if I've had friends

in need. 'Had a bad day? Here, have a warm loaf of bread.'
It's my version of the consoling cuppa.

I am not alone in my reverence of rolls and loaves.
Festivities around the world are celebrated by the baking of
special breads. Hot cross buns (British), *kulich* (Russian),
tsoureki (Greek), *panettone* (Italian), *Stollen* (German),
bakshalu (Indian), *himbasha* (Ethiopian) and *laakhamari*
(Nepalese), to name only a few, are all baked for special
occasions.

Bread is also religiously significant. In Christianity,
bread represents Christ's body. The Bible is loaded with
references to bread. In Jewish tradition, a special bread
known as *challah* is blessed and eaten during Shabbat meals;
it represents the manna that God provided for the Israelites
during their 40 years in the desert after leaving Egypt.

Somewhere along our cultural development, bread has
also become a symbol for procreation. The expression 'bun
in the oven' is rooted in older, deeper metaphors where the
mystery and magic of a rising dough has been compared to
the swelling of a woman's belly in pregnancy; women seem
to be equally skilled at turning 'seed' into bread or babies,
though one is clearly much harder work and less suited to
being smothered in peanut butter. More recently, in the
Urban Dictionary, bread has become a direct replacement
for the word sex, as in, 'I pulled a hottie last night and had
me some good bread.' I felt it necessary to include this
because if you're like me, you might actually think this
refers to some good home baking and inappropriately ask
to sample it. Awkward!

In fact, bread is so fundamental to many cultures, it has
infiltrated many of our everyday expressions. We use it to
refer to our livelihood (bread and butter), money (bread, or
dough), generosity (cast one's bread upon the waters), loyalty
(know which side one's bread is buttered), meals (breaking
bread) and innovation (greatest thing since sliced bread). In
Poland, the expression 'a bread roll with butter' means
something easily achieved. In Italy, if you say someone is 'as

good as bread' they are well-mannered. In Greece, you might shut someone up who has gone on and on about an old achievement by saying 'stale bread isn't edible'.

And yet, despite bread's cultural significance and delightful taste, there are elements to its nature that can prove hostile for some. More and more people are choosing to avoid gluten because they have a genuine allergy or intolerance, or they claim gluten just makes them feel bloated and sluggish. The gluten-free products in supermarkets have exploded in their diversity and popularity. Bread has also become the arch-enemy of dieters, struck off the list of any low-carb or Paleo diet follower as though all breads are equal. In a survey of 1,000 Australian women, one in four women said they love bread but avoid it because of weight concerns. And yet, we have had a relationship with gluten for thousands of years, so why is it that this relationship seems to have suddenly gone sour?

The dawn of dough
Much like the first discovery of cheese curds, we will never know exactly what inspired our early ancestors to crush grass seeds and mix the resulting powder with water to form a dough, let alone bake it over a fire. Like most great inventions, it probably started out as a bit of an accident. Yet, despite so many unknowns, this processing method was such a crucial step in our evolution as farming humans that it has generated considerable interest within the research community.

The earliest evidence of small-scale plant cultivation comes from Ohalo II, an archaeological site discovered in 1989 in Israel's Rift Valley, submerged in several metres of water in the Sea of Galilee (a freshwater lake). This excavation site is relatively unique because organic material that would normally have degraded has been preserved within the anaerobic conditions of the lake sediments. Conveniently, every now and then, the lake dries up enough to allow archaeologists to excavate the site – a rare positive amid the despair of serious droughts. The excavations revealed the remains of hundreds of different types of animals and plants as well as the remnants of six brush huts, hearths and a human grave. Carbon dating has estimated the site to be around 23,000 years old. Among the organic material are charred grains, including barley (*Hordeum vulgare* L.) and wheat (*Triticum monococcum* L. and *Triticum turgidum* L.).

Within one of the best preserved huts was a large flat stone, set on sand and supported by pebbles – considered to be a working surface. And, as one might expect in any messy kitchen, surrounding this stone were concentrations of various plant bits of both the medicinal and food variety. Researchers from the Smithsonian, Harvard University and the University of Haifa (Israel) analysed the large stone to determine whether it had been used to crush grains, and to do this they needed to look at starch residues and try to associate them with different plant species. They recovered 150 different types of starch grains from the stone and found that none of them were likely to be associated with the roots or tubers of any plants. The majority were from barley seed, with about 49 other types having the characteristic features of cereal grains. They concluded that the large stone was probably specialised for grinding grains.

One of the hearths at Ohalo II also stood out from the rest in that it was paved with stones. The researchers hypothesised that this was evidence of a basic oven; nomadic people today still use a layer of heated stones to bake bread dough, so it wasn't a gigantic leap. Along with the oven, five sickle blades made of flint were found on the site. When the blades were analysed, the sheen on their surface was consistent with the pattern left by cutting the siliceous stalks of grasses. Remnant grains, a grinding stone, a basic oven and ancient scythes: solid qualitative evidence that wild cereals were being processed and baked by hunter-gatherers about 10,000 years before the transition to an agricultural lifestyle. The tools suggested that these hunter-gatherers might have been doing some small-scale cultivation, a clear step on the path to farming.

Grinding the grains would have made the starch more nutritionally accessible by making the particles smaller and helping to remove some of the indigestible components. Starch is an important carbohydrate in our diet. It is essentially a whole lot of glucose molecules bonded to one another, and glucose is important fuel for the body and brain. Just as animals store glucose as glycogen, most green plants store their glucose as starch – it is essentially raiding the larder of the plant. By grinding and baking the starch in these grains, the early bakers would have increased the available energy by 56 to 72 per cent compared with eating uncooked grains. This is a significant dietary advantage.

However, studying ancient food and its preparation is a challenging task. After all, food is made to be eaten. And what isn't eaten degrades, except in a few very rare circumstances, such as Ohalo II. There is, therefore, a bit of a gap between the evidence at Ohalo and the next references to bread making. An analysis of residues in pottery recovered from two Libyan Saharan archaeological sites, dated from around 10,000 to 8,000 years ago, revealed the processing of grains in the pottery, but it is hypothesised that this was probably to obtain seed oil, although this doesn't rule out bread-making as well.

The first written reference to bread comes from the ancient Egyptian civilisation. Translations of hieroglyphs and artistic depictions suggest that wheat was coarsely milled and kneaded into bread dough that was rife with indigestible wheat husks (chaff). Reliefs in the Mastaba of Ti (a tomb in Saqqara) show bread in flower-pot-shaped moulds. It is thought that the moulds were heated in a fire first. The dough was then poured into the hot mould, covered, and left to bake by the heat of the mould alone. The artwork shows an evolution in Egyptian bread making over the next 1,500 years; the moulds became conical in shape and loaves are depicted baking in rudimentary ovens.

By the time of Ramesses III (approximately 1186 to 1155 BC), barrel-shaped ovens feature in the artwork, baking flat round loaves on the inside walls much as some breads are still baked today in places like Tunisia and India. But like all art, this is open to interpretation.

Luckily, arid desert conditions preserve organic materials just as well as lake sediments. The remains of about 70 loaves of bread have been recovered from two ancient Egyptian villages thought to have housed labourers, dating as far back as 3,500 years ago. Analysis of these loaves has found that bread of the time was diverse in terms of its texture, ranging from fine to coarse, though it usually had a rather dense crumb (crumb in this sense is a baker's term for all of the inside bit of the bread that's not crust, rather than the little bits left on a plate). The loaves were darker on top and lighter on the bottom, proving that they were indeed baked. The bread was almost always made of emmer wheat (*Triticum dicoccum* Schübl.), and sometimes other ingredients, such as coriander and fig, were found worked into the loaf. Under the scrutiny of a scanning electron microscope, the ancient bread looked much the same as a modern loaf. There was evidence of yeast, suggesting it was leavened bread, and there was a protein matrix and starch molecules, as expected. The structure suggests the dough probably wasn't kneaded heavily, but other than that, the samples looked just as 3,000-year-old bread should ... like bread.

Much like cheese, bread has a basic recipe that must be followed in order to achieve the final product: mixing flour, water, salt and, if leavened, yeast. This has not changed since these ancient times, but also, as with cheese, different cultures have tweaked this recipe to evolve a diversity of breads around the world.

More than anything else, it is the flour that determines the nature of the bread. Flour can be ground from grains, roots or tubers and breads of the world reflect what resources would have been readily available. For example, corn flour-based flat breads are common in the Americas as this is where maize was first cultivated. The breads are flat because corn lacks gluten, which provides the strong scaffolding necessary for a leavened bread. Rye and barley were more popular in northern climates because these cereals are more tolerant of the cold than wheat. These grains produce dark dense breads and usually require a good sourdough starter in order to rise. Many different Asian countries have rice flour-based breads, which, like cornbread, lack gluten and are therefore usually flat. There are also breads made with cassava root, sweet potato and pulses. If it can be ground into a powder, it can probably be made into some form of bread.

It is not only the type but also the coarseness of the flour that influences the texture of the bread. The Romans are credited with refining the milling process by using two grinding stones, one stationary and the other rotating. This created a finer flour that would develop into a spongier, more airy loaf.

The flour mixed with water makes the basic dough, but this can be enriched with milk, egg and sources of fat such as oil or lard. Sugar, honey or molasses can be added to sweeten the dough. Spices, herbs, seeds, nuts or fruit can also be added. The possibilities are endless.

The dough can have leavening agents added or not. It can be shaped into moulds, braided, rolled, stretched and stuffed. Finally, the bread dough can be baked, fried, steamed or boiled ... or combinations thereof. The end product might be tooth-crackingly dry so that it can be kept for long periods of time, or it might need to be eaten the moment it is pulled from the steamer in order to be at its most palatable, but most breads fall somewhere in between.

This is clearly the abbreviated version of bread-making. In order to truly get into the science behind bread in any detail, we must start with that bane of the modern diet: gluten.

The gluten matrix

A few years ago, while I was doing a week-long food course at the Massachusetts Institute of Technology (MIT) as part of a journalism fellowship, Dan Souza, the Senior Editor for *Cook's Illustrated*, came in to show us some of the science behind food. One of his demonstrations involved gluten. He had washed a ball of bread dough thoroughly to remove all the starch, leaving him with a shiny ball of pure gluten. To demonstrate to us just how elastic gluten is, he shaped it into a small circle and folded it over the nozzle of a compressed air tank. He turned on the air, creating a decent-sized gluten balloon. It was the ultimate demonstration of how this elastic substance traps air and helps create that irresistible bread structure so many of us love. And because gluten can literally make or break bread, the science behind its formation has been thoroughly studied.

Wheat flour contains two classes of proteins, called glutenins and gliadins, which make up about 10 to 15 per cent of the wheat flour. Gliadins are small proteins and a single variety of wheat may contain as many as 40 different gliadin proteins. In contrast, glutenins are protein mammoths formed of long chains of amino acids and weighing as much as 70 times more than a gliadin molecule. Sitting in a bag in your cupboard, these protein molecules don't do a whole lot. It isn't until water is added that gluten can start to form.

Neither of these protein classes are particularly soluble in water as they are dominated by hydrophobic amino acids like glutamine. But when water is added to the flour, it forms bonds with certain parts of the proteins, which forces the proteins to change shape: the glutenin molecules begin to uncoil. The more relaxed glutenin molecules begin to align and different bonds (disulphide, hydrogen and ionic) start to form between the protein chains. The gliadin proteins form non-covalent bonds with the aligning glutenin molecules, which helps the formation of further crosslinks and ultimately creates the protein matrix known as gluten. This matrix helps to trap starch granules and gases, creating the airy

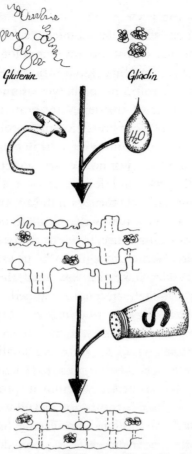

Figure 3.1 The proteins glutenin and gliadin become hydrated when water is added and start to stretch out with the help of physical force, forming the gluten matrix. Salt helps strengthen the matrix by neutralising charged regions on the gluten proteins.

pockets in a good bread crumb. As you would expect, the more gluten proteins present, the stronger the matrix, which is why high-protein flours are used for bread and pasta, while flour made from lower protein content wheat is reserved for pastries where a softer structure is more desirable.

Everything else we do to bread dough is to help encourage crosslinking of the proteins, thus building a stronger and more flexible gluten matrix. Finer flour, for example, has a greater surface area compared with coarse flour, making it easier to hydrate the protein molecules. Better hydration means better unfolding of the proteins and more crosslinks. Kneading helps in the hydration process by encouraging the protein molecules to unfold and align. If a dough is under-kneaded, fewer water molecules bind to the protein and fewer crosslinks are formed between the proteins, creating – you guessed it – a weak matrix. Alternatively, if it is over-kneaded, the bonds and crosslinks start to break down again and the matrix collapses. It's all about the matrix.

Salt is a relatively recent addition to bread as it was simply far too expensive in most countries prior to the eighteenth century. Salt obviously adds flavour to the bread and helps to slow yeast fermentation,* but it also helps to strengthen the gluten matrix, particularly of low protein content flours. Regions of the gluten proteins carry positive or negative charges, so when they are aligning and two like-charged regions come close, there is a natural repulsion. It would be like trying to line up two chains in parallel that had magnets haphazardly interspersed throughout. If magnets with the same polarity were to be brought into alignment, the chains would be pushed apart in this region and no longer be parallel. When salt is added to the flour-and-water mixture, it dissolves into its ions – sodium and chloride, which each have a positive and negative charge, respectively. These ions bind to oppositely charged regions of the gluten proteins, neutralising the charge and allowing the proteins to align more easily. Essentially, salt takes the magnets out of the chain.

Archaeological evidence suggests that by about 500 BC, microbes were regularly being added to bread dough to create leavened bread. Perhaps some ale or wine was added

---

* At this point, you are no doubt wondering why, if salt slows fermentation, it is added to pickles and sauerkraut and other fermented foods. In these cases, the fermentation is being carried out by lactic acid bacteria. Salt is added to stop less friendly bacteria from growing, while encouraging the salt-tolerant lactic acid bacteria.

to the bread dough for flavour, only to find that it made a lighter loaf. While the gluten matrix provides the scaffolding, it is ultimately gases that raise the roof of the loaf. This is where yeast comes in. When water is added to the flour, enzymes present in the wheat break the starch into smaller molecules. The yeast then uses its own enzymes to break these smaller molecules up even further, into glucose molecules, which it then ferments into carbon dioxide and ethanol. It is the carbon dioxide gas, trapped by the layers of gluten, that causes the bread to rise, while the ethanol evaporates during baking, creating lovely aromas. This fermentation step also introduces flavour compounds into the bread and the longer it is allowed to ferment, the more flavour is imparted to the dough. However, it is necessary to punch down the dough periodically (which should not be as violent an exercise as it sounds), otherwise large bubbles form.

These days we often shake in a bit of reliable easy-bake yeast. This is usually a combination of dried yeast, of the species *Saccharomyces cerevisiae*, and ascorbic acid (vitamin C). I'll come back to the ascorbic acid, which is a bread improver, later, but first I want to discuss the yeast. Before it was available in convenient dried packets and tins, yeasts had to be acquired from the environment. Food was left open to the air and wild yeasts settled on it and began fermenting. This seems straightforward for us with twenty-first century knowledge of microorganisms, but the organisms responsible for fermentation weren't discovered until the nineteenth century. The early bakers, brewers and winemakers knew only that adding a little bit of the fermented product to the new batch seemed to help speed fermentation along and produce an airy loaf of bread; a bit of yesterday's dough would be added to today's batch, for example, or the foam from fermenting ale was scraped and added to the bread dough. It is a method referred to as 'backslopping', but what they were in fact doing is domesticating strains of yeast that were particularly efficient at producing desirable end-products.

I must admit that yeast domestication sounds odd to me. It conjures up images of cowboys lassoing wild yeast cells and breaking them in until they are tame enough to throw into a bowl with flour. It's far less exciting than that, of course, but genetic analyses have confirmed that the thousands of commercial yeast strains used today stem from only a few ancestral strains, which through hundreds of years of backslopping have become genetically distinct from their wild counterparts. Humans have carried these yeasts on grape vines, in beer and in bread as we have migrated around the world and, like most things humans have tamed, these yeasts are now maladapted for life in the wild. Luckily for us, they thrive in vineyards, breweries and bakeries, performing the tasks they were bred to do.

The yeasts can't simply be left to their own devices in the bread, however. There is a rather delicate balance, and the mix of ingredients as well as the way the dough is handled helps keep it all in check. The fermentation must proceed along at a rate that doesn't outpace the development of the gluten network. If fermentation happens too rapidly, too much starch is fermented and too much gas is produced; the result is a weakly structured bread with big cavities. Sure it's soft, but your jam goes right through the holes! If insufficient fermentation happens, you get a very firm bread with little rise, which can be very crumbly, making spreading anything a challenging task.

The gluten matrix itself also needs to be handled with care. For example, once the dough has been shaped into its final cooking form, whether that's in a loaf tin, or formed into a round or tapered baton, it needs to be left to rest (prove). This gives the gluten matrix time to relax after being stretched so that when it is baked it stretches with the expansion of gases rather than tearing.

There is a reason there are entire websites dedicated to helping home-bakers analyse their misshapen, crumbly loaves; it's a delicate process and loaf forensics can be used to understand where things went wrong.

Reshaping rolls: from Romans to reducing agents

There is such skill in making bread that guilds and other associations were formed in order to protect the craft. As early as 168 BC, Roman bakers had formed the professional association known as *Collegium Pistorum* (the College of Bakers). They not only baked the bread, they milled the grain. This was a lifelong commitment as neither the bakers nor their children were allowed to withdraw from this *Collegium* in order to take up another trade – and few would want to, as the bakers were freemen who were highly respected and given special privileges. A representative of the college held a seat in the Roman Senate.

The writings of both Rome and Greece suggest that many different breads were made in these times: leavened and unleavened breads, made from rye, emmer, barley and other grains. But even as early as the third century AD written evidence suggests that white bread made from wheat was favoured and the darker wholegrain breads were left to the slaves and pigs.

Breads were marked with distinguishing imprints as they were often baked in communal ovens. A loaf of bread baked 2,000 years ago was preserved within the volcanic ashes of Mount Vesuvius and was found imprinted with the words 'Celer, slave of Quintus Granius Verus' – Celer being the baker. The imprint was made with a metal stamp, which helped to distinguish the loaves either as belonging to a particular family or as a product of a guild member or particular bakery. It may also have been marked with a stamp to identify it as being made for the purpose of free distribution to the people of Rome. Bread was added to the list of free foods doled out to keep the Roman people fed during the reign of Lucius Domitius Aurelianus Augustus (aka Aurelian) in the third century AD. The practice of marking breads with an identifying stamp would be picked up in other countries and was widely practised well into the nineteenth century here in Britain.

The first records of a Baker's Guild in England date back to 1155, making it the second oldest Guild in London next to the Weavers. They were highly respected and apprenticeships lasted seven years. But, because bread was such an important foodstuff, bakers and their products were also highly regulated. During the thirteenth century the first law on bread was passed, which regulated the price, weight and quality of bread manufactured and sold. It made the cost of a loaf standard and if the cost of wheat went up, then the size of the loaf was simply adjusted to be a little smaller so that the cost remained the same. Court sessions were held to deal with bakers caught cheating by overcharging or selling underweight or poor-quality bread, and the punishments usually included fines or disgrace in the town pillory. In London, the home of the Worshipful Company of Bakers, punishment was more severe. For a first offence, the baker was tied to a wooden panel (hurdle) and dragged from the Guildhall through the crowded streets to his house (talk about naming and shaming!). On a second offence, the hurdle's final destination was the pillory rather than his house. A third offence was the final strike; as well as the standard dragging through the streets, the baker's oven was smashed to pieces and he was never allowed to practise the art of baking again. The stamps on the bread made them completely traceable to a particular baker, as a strong disincentive against cheating. In fact, bakers were so scared of getting caught underweighting that they would throw in a small extra piece of bread, known as the 'in-bread', or an entire loaf if a dozen were purchased, hence a baker's dozen.

Similar regulation of bread did not happen in the colonies of America until 1646 when an Assize of Bread was enacted, which was modelled after English law. Identifying marks to make the bread entirely traceable back to the baker were required and there were two bread inspectors in every town to ensure the regulations were being followed.

From the late sixteenth century and throughout the seventeenth century, countries around the world were afflicted by famine, and bread – a diet staple – got creative. In Russia, there is evidence of breads being made at the time from birch and elm bark, straw, buckwheat husks, pigweed, acorns, potatoes, lentils, lime leaves and wild chervil (better known as cow parsley). Clay was sometimes added in for good measure. Most of these additions were horrible-tasting and indigestible, and certainly not nutritious. But such bread temporarily dulled the aches of empty bellies, even though it probably introduced some new discomforts. The proportion of adulterants would shift with the availability of grains, waxing and waning as needed. It would have been considered fraudulent had people been trying to pass it off as a pure grain bread, but everyone knew this was famine food.

In 1756, British wheat crops failed. Instead of preparing to bring in the harvest, farmers were forced to sit idle, watching as torrential rains pummelled their wheat into the ground. In response, the British government dictated that a standard bread, marked with an 'S', should be made, which had a higher bran content and therefore provided more nourishment per gram. 'Low grind' flour was used, which was passed through the millstone only once so that it contained all of the original components of the wheat, and therefore all of the nutrients. 'Higher grind' flour, by contrast, was passed multiple times through the millstone and then sifted to remove the bran, thereby losing a lot of the nutrients. The government's move made sense: if there is a limited supply of wheat, don't waste anything and get as much nutrition from it as possible. Good in theory, except this wasn't acceptable to the British public. The love of white bread documented from Ancient Greece and Rome was alive and well in eighteenth-century Britain. Books were judged by their covers and people were judged by the bread they ate. Those who could still afford it would go to great lengths to seek out white loaves of bread and undoubtedly position it prominently at the top of the shopping basket while strolling down the high street.

Of course, whenever times are tough and there is a premium product to exploit, there is an opportunity for fraudulent activity. British bread was not immune. With

wheat in short supply and the wealthy still wanting white
bread, bakers began to purchase inferior white flour, which
produced grey loaves. To counter this, they mixed in alum
to make it white. Alum, which is potassium aluminium
sulphate, has been used since medieval times as a fixing
agent for dye. It was also used to preserve food, make
pickles crisp, and, as it turns out, make inferior bread white.
But alum is not without its effects. While the poorest
people without the resources or will to dine outside their
'bread class' ate wholesome brown bread, those trying to
keep up appearances were struck with abdominal cramps,
nausea and even vomiting as their guts tried to deal with
the toxic effects of the alum. As the price of wheat went up,
so did the proportion of alum in each loaf.

Alum use peaked during the crop failures of the
mid-eighteenth century, but it would continue to be an
adulterant of bread for at least another 100 years, despite its
ban in Britain in 1758. This deceit even made its way across
the Atlantic to the US. An article in an 1848 issue of
Scientific American (vol. 4 (5): 38) describes the use of alum
in English bread, stating that 'the custom is so universal
that the most respectable baking establishments are stained
with the crime'. It goes on to describe to the US readership
how repeated consumption of alum hardens the bowels
and causes constipation, which is then countered through
the use of laxative medicines – dealing with the symptoms
rather than the cause. The unnamed author of the article
claims that in this way 'the consumer is first robbed by the
baker of his money and his health, and then again fleeced
by the quack'. However, determined to leave readers of the
article empowered, the author provides a simple at-home
test to determine whether bread has been adulterated with
alum. First, soak a sample of the loaf in water. To this
water, add a solution of muriate of lime (formed from
calcium-containing lime and hydrochloric acid), which
presumably would have been fairly accessible in this
pre-eBay world as it was used in medicine, water treatment

and concrete development. If any alum is present, the water will turn milky white, otherwise it will remain clear. The author's explanation of the chemistry is a truly colourful description of a chemical reaction: he states that the milkiness is formed because the sulphuric acid in the alum is so overwhelmed by its attraction to lime that it quits its association with the other elements in order to form a new relationship, known as sulphate of lime (calcium sulphate). This is how I would imagine Jane Austen would write about science. The introduction of adulteration laws and enforcement in both the US and Britain, as well as wider awareness among the public, eventually helped get alum out of the bread. By 1877 only 7.4 per cent of bread tested in Britain was adulterated, and by 1888 this was down to less than 1 per cent.

So, by the end of the nineteenth century, bread was once again a pure product and still a staple of most households. Manufacturing practices had changed for most industries with the industrial revolution and yet grains, for the most part, were still being milled using millstones driven by wind or water power. While this wasn't the most efficient process, there were some advantages to the flour produced. The stones grind slowly, heating the grain gently without making it too hot – excessive heat can compromise the fat and vitamins in the germ layer of the grain. The coarser grind also limits the surface area exposed to oxidation and therefore degradation of the flour. Many bakers today still prefer the taste and behaviour of a stone-ground flour.

Yet urbanisation and industrialisation begged for a more efficient flour-milling process and millers had begun to attempt greater efficiency over the last 100 years. In 1786 a steam engine was introduced to drive the flour mill in Battersea (a district of south-west London). It was so efficient that it ground the same amount of flour in one year as all of the other flour mills in London combined. It mysteriously burned down after five years. Then, steel-roller mills, which were already in use for grinding animal feed and for malt milling, started to make their way into flour milling. They were first introduced into Hungary and Switzerland as early as 1820. For a while the rollers were used in combination with millstones: the steel rollers broke open the grain kernel, while the stones would grind it. There is some disagreement among the sources as to who perfected the roller mills, so I will just say that the system went through numerous iterations. In the end, it involved a set of toothed steel rollers used to break the wheat, with a set of smooth-surfaced steel rollers below to crush it into flour.

American millers soon adopted the steel-roller system and improved upon it by adding mechanical sifters and purifiers, which easily removed the germ and bran to produce a lovely white flour. The Brits were a little slower on the uptake. By 1887, only 5 per cent of mills in the UK were fully converted to roller mills and yet these mills were producing 65 per cent of the country's flour. But by the turn of the century, Britain was importing more wheat from Canada and the US. Although this harder wheat was better for bread, due to its high protein content, it was also tough on the stone mills. Britain was forced to convert to roller mills. By the start of the nineteenth century, many mills were state-of-the-art facilities with whole areas dedicated to preparing, washing and drying the wheat before sending it to the milling area, where more than 40 rollers, all driven by the power of steam and of various sizes and rugosity, would break the wheat before it was sifted, purified and sorted for further grinding. This is known as a reduction roller system and it churned out about 70 sacks of flour (8.6 tonnes) an hour, double the capacity of the state-of-the-art mills from only 20 years earlier. And the flour was whiter than any unadulterated flour before it. White flour and therefore white bread was now affordable for all and no longer just for the financial elite. In response, the rich had to show off with quantity instead. In 1900, an upper-middle-class family could go through as many as 24 loaves of bread in a week to supply the endless dinner parties and afternoon teas they hosted in order to affirm and even improve their social status.

The milling process had become incredibly efficient at removing all of the bran and germ from the wheat, which incidentally is also where a significant portion of the nutrients and fibre are found. The bran, which is the outer layers of skin of the wheat kernel, contains important antioxidants, B vitamins and fibre. The germ is the part of the kernel that sprouts into a plant – its embryo – which contains B vitamins, protein, minerals and healthy fats. All of this was being removed, and to add to this, millers began adding flour-bleaching agents to speed up the natural ageing process of flour. Aged flour develops a stronger gluten matrix and incidentally, through oxidation, becomes whiter. But this process takes months. To condense three months' worth of ageing into a 48-hour period, millers began to add nitrogen dioxide, chlorine dioxide, chlorine and peroxides to the flour. This compromised the nutritional quality even further. Roller-milled bleached flour contained less than half of the calcium and a quarter

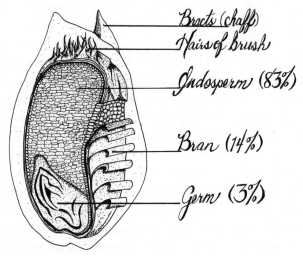

Figure 3.2 A wheat kernel.

of the iron per 100g compared with stone-milled white flour at the time. It also had less than 15 per cent of the vitamin B1 and none of the vitamin A. However, because the roller-milled flour was finer, it made more of the sugars accessible, so it had about 13 per cent more carbohydrate than stone-milled white flour, but similar levels of protein and fat.

Some people started to raise concerns about this loss of nutrition. In 1924, Dr Charles Shelly wrote in the *British Medical Journal* advocating changes to the regulations to ensure the 'absence of the vitamin-destroying chlorine and other additives'. Dr Harvey Wiley, Chief Chemist in the US Department of Agriculture (USDA) from 1882 to 1912, also campaigned strongly against the use of bleaching agents in flour. These advocates were preceded by the likes of Sylvester Graham a whole century earlier, who hated seeing the nutrients contained in germ and bran discarded; indeed 'graham flour' (a wholewheat flour mainly found in the US and Canada) is named after Sylvester. Yet, despite centuries of protests, efficiency and consumer preferences won out over nutrition and white bread remained popular; by 1930 America's Wonder Bread, a whiter-than-white pre-sliced loaf, was proudly displayed on kitchen counters across the US.

By the time the 1940s rolled around, a couple of periods of rationing had brought nutrient deficiency to the forefront of government agendas on both sides of the Atlantic. To add to this, diseases caused by vitamin deficiencies had reached worrying levels. Pellagra was first documented in the US in 1902. It is the result of a niacin (vitamin B3) deficiency and between 1906 and 1940 approximately 7,000 deaths a year were attributed to this disease. Most of these were concentrated in the poorer southern states where diets may not have been as well balanced. Governments around the world began implementing mandatory enrichment of white flour with

B vitamins (riboflavin, niacin and thiamine), plus iron and sometimes calcium. This enrichment is credited with the eradication of pellagra as well as beriberi, which is caused by a deficiency in thiamine. In the late 1990s, many governments added folic acid to the enrichment formula and subsequently saw significant declines in neural-tube defects in newborns – by 46 per cent in Canada and 28 per cent in the US, for example. The UK opted to advise pregnant women to take supplements rather than adding folic acid to their flour-enrichment formula, and it has been estimated that around 2,000 cases of serious birth defects could have been prevented since 1998 had the UK government chosen differently. Today, there are 86 countries globally that have mandatory wheat fortification programmes. It took them nearly four decades to start putting back what the steel roller mills had become so efficient at removing: nutrition.

Commercial bread makers started to look towards other areas of the process to gain efficiencies and the long fermentation time was the obvious place to start. Racks of dough and tinned loaves took up valuable space on the floor of commercial bakeries and hours were spent waiting for yeasts to expel their gases. Considerable research was put into finding ways to reduce this fermentation time and in the early 1960s, bread-making began to change considerably.

In the UK, the Chorleywood Bread Process (CBP) was introduced in 1961. It was developed by the Flour Milling and Baking Research Association at Chorleywood, a village north-west of London. The process relies on a lot of yeast (proportionally twice as much as I use in my bread) and brute force to speed the dough development along. The ingredients are mixed in a high-speed mixer for anywhere between two and five minutes. This short but intense mixing time is also when the dough is developed. The mixer transfers energy to the dough, heating it up while also mechanically changing the conformation of the wheat proteins so that the gluten matrix forms quickly. Air bubbles are trapped within the dough during this vigorous exercise to add to the gases being produced by the activated yeast. The mixer is kept in a partial vacuum in order to control the size of the gas bubbles. Each batch that is mixed can produce 350 loaves of bread and they can mix 20 of these batches in one hour. However, in order to produce a reliable loaf consistently, a couple of additional ingredients – beyond those added to the flour at the mill – also have to be added to make this process work.

First, an oxidising agent is added into the mix. Oxidising agents act in several different ways at the level of bond formation between the gluten proteins. Without getting into the nitty-gritty of disulphide bonds (the mere mention of which brings back vivid memories of chemistry lectures that left my head spinning), the oxidising agent pulls the most reactive parts of the proteins away. In doing so, it encourages more bonding between the protein molecules, which leads to a stronger, more elastic gluten matrix that will stretch when gas forms in the early stage of the cooking process, rather than breaking. The dough and thus the bread is, therefore, 'improved'. All sorts of chemicals have been added as oxidising agents over the decades. Potassium bromate was a favourite for a while as it was economical. However, by the 1990s it was listed as a possible carcinogen to humans, which led to its ban as a food additive in the EU, Canada, China, Brazil and a number of other countries. However, if used in the correct quantities and baked for the correct amount of time, the bromate should be completely decomposed and it is on this premise that the US FDA still permits bromated flour, though it encourages bakers to avoid it. These days, ascorbic acid (vitamin C) is the most commonly used bread improver. Interestingly, it must be listed on the label as ascorbic acid rather than vitamin C because it is being used for its functional properties rather than its nutritive value.

Fat and/or an emulsifier is the other ingredient that is necessary to add as part of the CBP. The addition of fat to bread is not new. Historically, lard has been added to white bread to add flavour to what would otherwise be an incredibly flavourless platform for spreads and preserves. Fats make the bread softer and admittedly I will add a splash of olive oil to mine from time to time. However, high-melting-point fats (animal fats) or emulsifiers are necessary in rapidly prepared doughs in order to stabilise the loaf during the final proving stage. Without this addition, bread prepared by CBP or similarly fast processes has a blistered crust, a firm crumb and lacks spring. It was through experimentation that the right ratio of fat was agreed upon when developing the CBP, ranging from 0.1 per cent of flour weight to as much as 4 per cent for a wholemeal. Considerable research has gone into perfecting exactly the type of fat that goes into these fast-production breads, so that it creates the ideal loaf with the addition of as little solid fat as possible. Coconut oil and palm oil are both commonly used now, though the former is expensive and the latter is an environmental nightmare. Fat also helps to extend the shelf life of the bread slightly by slowing the movement of free water in the bread as it naturally migrates towards starch crystals that form over time.

Once mixed, the dough is cut into specific single-loaf weights, and it has 30 seconds rest as it moves along a conveyor belt, where it is then shaped. The dough is flattened into a pancake and rolled like a Swiss roll, then it is cut into four, and these four Swiss-roll-like shapes are stacked into the bread tin. Instead of using time to create texture, layering the dough like this accomplishes the same thing. The loaves are left to prove for an hour and there has been no way to shorten this step as yet. In a big commercial bakery about 7,000 loaves are likely to be proving at any one time. The loaves are then baked for 20 minutes and cooled for an hour before they are sliced and bagged. The whole process from flour to wrapped pre-sliced loaf is about three and a half hours.

The Chorleywood Bread Process marked a revolution in bread making in the UK. As well as significantly reducing the time needed to develop the dough, CBP also allowed the use of softer wheat varieties, which meant that more British wheat could be used and bakeries didn't need to rely so heavily on imported North American wheat. Around 80 per cent of bread in the UK is still made using this method. The process has also been adopted in Australia, New Zealand and India.

Across in the US another process for commercial bread production was being developed in the early 1960s, known as activated dough development (ADD). The acronym is very appropriate because it is ADDitives that are key to this process. Unlike the CBP, which relies on mechanical force to move dough development along, ADD uses chemicals. It requires the addition of extra yeast, oxidising agents and a fat or emulsifier, for the same reasons as discussed above with the CBP. In addition to this, however, it uses l-cysteine hydrochloride, which is a salt that precipitates when l-cysteine (an amino acid) and hydrochloric acid are combined. L-cysteine acts as a reducing agent, losing electrons and helping to form linkages between the gluten proteins to make an airy loaf. L-cysteine is manufactured by the hydrolysis of poultry feathers and animal hair, and it is widely thought that human hair was used as a main source as well. I couldn't find any reputable reference to this, but the EU does explicitly state in its food additive guidelines that 'human hair may not be used as a source for this substance', suggesting that this is at least a possibility. It can also be synthesised using microbial fermentation to eliminate dependency on animal products, but this is a more expensive source. The ADD method was widely adopted because it didn't require bakeries to purchase expensive high-speed mixers – they could use whatever low-speed equipment they already had. As a result, small bakeries in the UK also started to adopt this method. By the addition of a few extra ingredients (some extra water was also necessary) the fermentation time in bread making was slashed down to about 30 minutes. Chemically enhanced bread filled bakery shelves everywhere for a couple of decades. But then some of these additives, such as bromate, were banned and consumer attitudes towards additives changed. Or perhaps they just became more aware that they existed in the first place. ADD is still used by small bakeries, but most commercial bakeries in North America are using a sponge-and-dough method where part of the flour is mixed with water and yeast and left to ferment and then later added to the rest of the ingredients. Additives are still used to improve the quality of the bread and extend its shelf life.

Just as with cheese, commercial bakers looked to those natural biological catalysts – enzymes – to help speed things along and improve profit margins. While the use of enzymes was already in practice to some degree, the deregulation of enzymes in Britain in the mid-1990s opened up new opportunities. Wheat naturally contains enzymes that break starch into maltose, as mentioned previously, but one of these enzymes, known as alpha-amylase, is in small quantities and therefore limits the rate at which starch can be broken down. Millers often add a bit of fungal alpha-amylase to the flour. The fungal source of this enzyme is less heat-tolerant and therefore gets inactivated early in the process, after it kicks off the fermentation process and before it starts breaking the dough down.

This enzyme may also be added at the time of mixing, particularly when using softer wheat varieties as is often the case with facilities using the CBP method. In this case, it's added to the dough because it has been found that it lets the dough continue to expand in the beginning part of the baking process more than it would without it, resulting in a much springier, fluffier loaf.

Science has of course enabled detailed knowledge of these enzymes and with that comes the ability (and desire) to manipulate them. Enzyme manufacturers have now been able to identify and select naturally occurring enzymes that produce certain desirable products as starch is broken down, but they still have that all-important low inactivation temperature. The result is that it is now possible to control the breakdown products made throughout the baking process, building in desirable traits such as a softer crumb that doesn't go stale as quickly.

Enzymes other than the amylases have found their way into the bread-making process as well. In North America, proteinases are sometimes added to help counter the overly strong gluten structure formed from hard North American wheats; this is rarely an issue with the weaker wheat grown in Europe. Lipases, which help to break up fats, are added to produce molecules composed of a glycerol and a single fatty acid, known as monoglycerides, which help keep the bread moist for longer. Asparaginase is added not only to bread but also to other baked goods, as it converts the amino acid asparagine into another common amino acid, aspartic acid. Asparagine, which is naturally present in starch, forms acrylamide when heated. This is responsible for the brown crust and toasty aroma and flavour of our favourite baked goods, but it is also a suspected carcinogen. By using asparaginase to convert asparagine into aspartic acid, acrylamide production can be reduced by as much as 95 per cent. As enzymes are naturally occurring, the main concerns around their use are for those in the bakeries that are handling them. All enzymes are proteins, so there is a potential allergenic aspect, particularly as they might get into the air and then come in widespread contact with the skin and respiratory system. Bakeries that use enzymes must take specific measures to minimise exposure for their employees.

Preservatives are the other addition to most commercially produced breads. This is for one main reason: bread is travelling more than it ever has in the past. In Canada, there seem to be roughly 100 commercial bakeries and another 1,100 or so retail bakeries for the whole of the country; bread is travelling as much as 1,000km (nearly 700 miles) from bakery to point of retail. In the US, there are approximately 2,800 commercial bakeries and another 6,000 retail bakeries. In the UK, just three commercial bakeries – Allied Bakeries, Hovis and Warburton's – hold three-quarters of the market share. Compare this with France, which is only just over 6 per cent of the area of the US (it would fit within the state of Texas) and has 20 per cent of its population, yet has over 32,000 artisanal bakeries. As I sit here in France buying fresh bread daily, I can't help but think that they have got it right.

However, if you have to move bread great distances then it needs some help to prevent it from going stale and mouldy on its journey. Sorbic acid is a commonly used preservative in the food and beverage industry that is particularly effective at inhibiting the growth of fungus. This is wonderful, except that one fungus, yeast, is clearly a rather critical component of bread. In non-leavened breads, sorbic acid can be added readily and can increase the life of the bread by 100 to 500 per cent. If sorbic acid is used with any leavened dough, though, it has to be sprayed on after baking to inhibit yeast growth on the surface of the bread. The salts of propionic acid (a naturally occurring acid that smells of body odour) , such as calcium propionate, are added into many baked products as another effective preservative. As with all additives, there are limits to how much can be used, but within the legal limits set by the EU, calcium propionate can extend the shelf life of a loaf by two to three days. Vinegar or acetic acid also has preservative qualities and sounds a lot nicer on an ingredients label, but it isn't nearly as effective at inhibiting mould growth.

There are other additives that you might see on the ingredients list of commercial loaves. Soya flour is often added to help produce a softer crumb. Fermented wheat flour is added to improve the flavour; a portion of dough is fermented for a long period of time, developing flavour compounds, and so adding this 'sponge' to a rapidly made dough provides a depth of flavour without having to ferment the entire batch. Sugar, in various forms, including high-fructose corn syrup, molasses and honey, is also added to help speed up the action of the yeast and reduce fermentation time. Sugar is much more commonly added to wholemeal bread to mask the more bitter taste of bran, ironically in an attempt to encourage consumers to choose the healthier wholemeal option.

Improvers, emulsifiers, fats, enzymes, sugar and preservatives: all of these additions to bread dough are finely tuned to churn out the best bread possible in the least amount of time. They can be used to help counter minor inadequacies in the wheat and to help compensate for shortcomings in the processing methods, such as the severely reduced fermentation time and the efficient removal of nutrients. The ingredients list of bread has doubled in order to compensate for decades of tinkering with the bread-making process, but have we gone too far?

Clearly the addition of nutrients to flour has helped to counter some horrific diseases. And yet, had we left those nutrients in the flour to begin with, these diseases would probably not have presented themselves in the first place. But the attractiveness of a white fluffy loaf of bread is undeniable, even for a wholemeal, granary girl like myself! Have you ever held a piece of white bread in your mouth for any length of time? It's like sugar. No, wait, it IS sugar. As the enzymes in your mouth break down the undigested starch in the bread, it's converted to sugar. With wholemeal bread, that sweetness is countered by the bitterness of the bran. It's hard to deny a couple of million years of evolutionary history that drives us to go for the easy, fast energy source – white bread!

There are also benefits to some of the other additives. If you recall from the introductory chapters, I mentioned white flour tortillas and the fact that glycerol is added as a humectant in order to keep the wrap soft and supple. Let me tell you what happens when this isn't added. My son loves wraps for his packed lunch. Keen on reducing the amount of processed food I bring into the house, I thought I could make flour tortillas fresh each morning for his lunch. I can only explain this as a moment of temporary insanity. I whipped up some tortillas one morning, cooked them in a hot pan and then wrapped the warm floury flat bread around some cheese before popping it into his lunchbox. *I might actually be supermum*, I thought to myself. Proud as Punch, I saved the extra tortilla for my own lunch. By the time I went to eat it four hours later, it was hard, brittle and thoroughly unpleasant. My son came home from school unimpressed. So, there are times when I must accept the benefits of additives.

Usually, however, we aren't terribly keen on additives, and this is changing our long-term relationship with bread. It appears to be on a fast descent from dietary staple to second-rate grub. This is, in part, due to the perception that bread is making us fat and making us sick. But in every relationship gone sour, it's always easier to blame the other party.

It's not you, bagel, it's me

Let's begin with the fat issue, and I'm not going to be gentle about this. Here's the scenario. An overweight man gets into his car and drives four blocks to his local pub (it could equally be a woman, but this is inspired by something I saw recently, so I'm going with it). He goes in and orders a burger and a pint and says 'no bun'. He sits down and enjoys his food while watching the football game on the big screen. He then gets back in his car, drives four blocks home and puts his feet up for a night of TV and maybe another beer ... or two. What is wrong with this picture?

Avoiding bread while consuming several units of alcohol and leading a sedentary lifestyle seems rather pointless. Bread is not the smoking gun of obesity, despite our desperation to find one. In the last 30 years or so, there have been no less than 38 studies that have examined the link between obesity and bread (as part of a broader look at diet). A review that looked at all of these studies came to these conclusions: 1) people who include bread in their diet are not any fatter than those who don't; 2) whole-grain bread is better than refined (white) bread; and 3) there may be a link between refined bread and excess abdominal fat. The authors of the review quite correctly point out that it's too complicated to sort through all of the lifestyle factors at play in these studies – people who choose wholemeal bread might also eat more fruit and vegetables, cycle to work and do hot yoga. Who knows? A study conducted in Norway of over 50,000 people, for example, showed that abdominal obesity was associated with people who ate less bread, particularly less whole-grain bread. But these more obese people also ate fewer vegetables and more hamburgers. They drank sugar-free soft drinks, skipped meals and snacked late at night. Humans are complex animals with diverse habits.

The evidence does, however, seem to be pretty clear about the nutritional benefits of wholemeal bread over white, highly refined bread. To remove some of the complexity associated with human subjects, we have of course spent many decades feeding rats all sorts of different human foods in order to watch their little rat waistlines. Bread is no exception. In 1970 a researcher from Texas fed one group of freshly weaned rats on enriched white bread and a second group on wholemeal bread. Within three months, over 60 per cent of the white bread rats were dead and those that weren't dead had stunted growth. The wholemeal rats, on the other hand, were leading a perfectly healthy lab rat (an oxymoron) existence. White bread, in our house, is a rare homemade treat that is usually served for purposes other than nutritional value; there is simply no denying that when it comes to French toast, a thick white bread just works better.

There is no evidence to suggest that bread is causing obesity. Bread isn't making us fat, we're making us fat! So if you are one of the many thousands of people who are depriving themselves of bread because of this belief, it is time to let it go. Keeping in mind, of course, that not all breads are created equal. If, however, you are avoiding bread because it genuinely makes you ill, then that is a different issue all together.

Coeliac disease is an autoimmune disease in which genetically pre-disposed individuals suffer an inflammation of the small intestine. The body's immune system is triggered by the presence of gluten proteins, mistakenly perceiving them as harmful foreign bodies and attacking them. This damages the lining of the small intestine and inhibits the absorption of nutrients. If it's not controlled, the surface area of the small intestine is permanently compromised and other complications begin to manifest themselves, such as iron deficiency, osteoporosis and malnutrition.

Over the last few decades, many studies have been conducted to understand the incidence of this disease. A UK study back in 1950 established that 1 in 8,000 people in England and Wales had coeliac disease, while 1 in 4,000 people in Scotland suffered. At this time the diagnosis could only be made based on people's descriptions of symptoms and so the disease was vastly under-diagnosed. Today, antibodies in the blood and a biopsy from the lining of the small intestine are used to make a diagnosis. These more advanced tools have led to estimates that 1 in 100 people in the UK have coeliac disease; even if they don't present gastrointestinal symptoms, they might still not be absorbing nutrients properly and suffer from issues such as low iron levels. This level of 1 per cent is in keeping with the prevalence of the disease globally, though there are most certainly regional differences. Prevalence in Germany is only 0.3 per cent, while in Finland it is 2.4 per cent and among the Saharawi culture of North Africa, prevalence can be as high as 5 per cent.

These geographical differences are in part due to inherited genes that are passed on within these populations, but there are diet differences as well. It has been claimed that the prevalence of the disease roughly correlates with wheat consumption in these different regions. This appears to be true in some cases. The average annual consumption of wheat flour globally is just under 70kg (154lb) per capita. North Africa (Egypt and Algeria) consumes on average 201kg (443lb) of wheat flour per person per year – more than double the global average. Germany, with its below-average prevalence of coeliac disease, also consumes a below-average amount of wheat flour per capita at about 61.5kg (136lb). And yet the Finnish, despite a greater prevalence of this disease, are only consuming about 46kg (101lb) of wheat flour each annually. Turkish people munch through over 220kg (485lb) each of wheat flour every year and yet the prevalence of coeliac disease is in keeping with the global average at just less than 1 per cent. Wheat might be the trigger, but it is certainly not the only factor explaining prevalence.

To add to this complex story, things can change over time for individuals. In one study, nearly 49 per cent of Finnish children who had tested positive for antibodies indicating coeliac disease were negative for these same antibodies three years later, despite continuing to consume gluten in their diet. There are recorded cases where people who have been diagnosed with coeliac disease in childhood later reintroduce gluten into their diets as adults and yet have no antibody response or change in their small intestine lining. However, it would also seem that people can get well into adulthood before they start to have a response to gluten – their tolerance changes.

There is definitely a genetic predisposition to the disease and gluten is without a doubt the trigger, but the variable natural history of the disease makes it difficult to understand it entirely. Gut microbiota, for example, might also play a role, but it's unclear what this role might be. Microbe species seem to be out of sorts in children with coeliac disease; some species are far more abundant than in normal patients, while levels of *Bifidobacterium*, a well-known beneficial group, are found in much lower levels. But which came first? The disease, or the shift in microbe communities?

Wheat wasn't identified as the trigger for the disease until 1944. A wheat shortage during the Dutch famine meant that wheat was all but eliminated out of the diet of patients in hospital for gastrointestinal disorders. The paediatrician Willem Dick noticed that his patients with coeliac disease improved during this time but then relapsed when bread was reintroduced to their diets. The trigger was clear.

Other people report gluten sensitivity and have symptoms that range from being tired and bloated to severe gastrointestinal discomfort – the latter symptoms being indistinguishable from coeliac disease. It's not known what the relationship is between this sensitivity and the disease, but people with a sensitivity don't develop autoantibodies or experience changes to the lining of their small intestine. Nor do they have any of the longer-term complications associated with nutrient malabsorption, which can develop with coeliac disease. The treatment for all of these cases is to remove gluten from the diet permanently. Some people have done this because they've heard how gluten can affect others, without any medical justification for avoiding it themselves. But are more people actually intolerant of gluten or are doctors just better at diagnosing it?

The answer is likely to be a bit of both. In the past, children with coeliac disease would have probably died either of dehydration with severe diarrhoea or of the complications associated with malnutrition. The cause of death may have been listed as gastroenteritis or failure to thrive. Now, thankfully, these children live, which means the genes associated with coeliac disease get passed on and become more frequent in society; the human gene pool has changed.

The wheat we are eating has also changed dramatically. When our ancestors first started to cultivate wheat there would have been significant genetic diversity, producing varieties with variable levels of protein. From that diversity, farmers would have selected those plants that were easiest to cultivate – they would have produced lots of seed which stayed in the seed head so that it was easier to harvest and it would have thrived in the local growing conditions. Our farming ancestors moved wheat around and bred it with other varieties. The wild diploid wheat (two sets of chromosomes) became tetraploid (four sets of chromosomes) and then eventually hexaploid (six copies of its seven chromosomes) in the farmer's fields. The bread wheat (*Triticum aestivum*) we grow today has about five times as much DNA as humans.

In 2012, the bread wheat genome was analysed, providing some insight into its ancestry and evolution since we began cultivating it. Bread wheat comes from the hybridisation of cultivated wheat (*Triticum dicoccoides*) and goat grass (*Aegilops tauschii*) around 8,000 years ago. Since then, bread wheat has been evolving rapidly, expanding copies of genes that have made it grow faster, resist disease and produce more seed. Having multiple copies of genes provided distinct advantages and these plants outcompeted any diploid relatives, particularly with helping agricultural hands. Then in the last few hundred years, varieties that had more protein content – the harder wheats – were selected out for their gluten-forming qualities and ease of processing. The result is that today wheat is less diverse and specifically selected for particular traits, including protein content and disease resistance. The wheat gene pool has also changed (and will continue to do so).

There is also evidence to suggest that how we treat the wheat may be contributing to our reaction to it. A 2013 study led by MIT linked the increasing incidence of coeliac disease not to better diagnostics, but rather to an increase in the use of the weedkiller Roundup on wheat (and other crops), and more specifically, Roundup's active ingredient glyphosate. Any fish exposed to glyphosate experience coeliac-like symptoms because the glyphosate prevents digestive enzymes from breaking macronutrients down, including proteins. The MIT group proposes that the glyphosate residues that have been found in the Western diet are the real cause for the rise in gluten intolerance. Glyphosate is sometimes used to dry the wheat (as well as oats and lentils and peas) to make it easier to harvest. It will be interesting to see what happens as countries begin to ban glyphosate.

Finally, the way bread is made has changed. There is some research out there that suggests that sourdough breads and breads fermented with traditional long fermentation periods promote healthier gut flora than rapid dough production methods such as the CBP. In patients with irritable bowel syndrome (IBS), sourdough produces less gas and other symptoms of IBS. It is thought that the lactic acid bacteria in sourdough, if given enough time, can break down aspects of the gluten that cause IBS symptoms and might even make gluten tolerable to some coeliac sufferers.

It has also been suggested that the bloating and discomforts that some people experience after eating bread aren't a result of gluten at all but rather of FODMAPs, which are short-chain carbohydrates and sugar alcohols that are poorly absorbed by the small intestine and therefore pass through to the large intestine where they are fermented by gut microbes. This fermentation leads to gas production and retention of water in the bowel, which can cause bloating, abdominal pain, constipation or diarrhoea ... or combinations of all of these things. Wheat naturally contains FODMAPs, and although I couldn't find any literature that looks at the FODMAPs associated with different bread-making processes, it isn't unreasonable to think that more fermentation outside of the gut by bacteria and yeast during the bread-making process would leave fewer FODMAPs for the gut flora to handle. But this is merely conjecture.

In 2015, 50 million fewer pre-sliced loaves were sold in the UK than the previous year. Bread sales in the US have gone down in volume, though increases in prices have helped offset this decline in terms of value. But consumers haven't given up on this relationship yet – we are simply shifting away from bland pre-sliced commercial loaves (particularly white) to speciality and artisan breads. We want granary loaves and multigrains. We're trying rye and spelt flour for a change. And sales of sourdough, with its reputation of digestibility and saliva-stimulating lactic tang, are going through the roof. As consumers, we don't want this 8,000-year-old relationship with bread to end, we're just demanding a bit more of it. Unlike our ancient ancestors, we aren't desperate for calories (though one could argue that we are still desperate for nutrition), and our purchasing decisions should reflect this.

However, it is hard to make purchasing decisions when the information we receive as consumers can be misleading. Not all of the additives (enzymes, *etc.*) that I have mentioned need to be listed in the ingredients list of bread. Some are considered 'processing agents', which may be added to the food during processing and are largely used up during that process so that they are no longer present in the final product in any significant amount. Some people may be perfectly content not to know what these are, while others, like myself, may feel that this secrecy contributes to our distrust of the food industry. There's also this ruse that we all fall victim to when we smell baking bread at the supermarket – that this is fresh bread that has been made on site. The reality is more likely that the bread was made by the same people who make those pre-packaged pre-sliced loaves and partially baked before it was flash frozen. The frozen loaves may be stored for as much as a year before being brought into a supermarket 'bakery' to be finished off in their ovens and sold as 'freshly baked' bread. It is also unclear when buying a sourdough whether the bread has had a long fermentation time or whether that sourdough taste has been achieved through the use of enzymes and other additives.

The French have made this very simple. In 1993, France enacted legislation that states that bread can only be sold under certain names (essentially variations of traditional French bread), if it contains no artificial additives, is made only of bread wheat flour, drinkable water and cooking salt, and if it contains bakers' yeast and/or leaven made of specific quantities of bean meal, soybean meal or malted wheat flour. The dough or bread can't be frozen at any point in the process. It can only be sold or marketed as being from that bakery if it is fully kneaded, shaped and baked in that establishment. In other words, it can only be called real bread if it is in fact what most of us think of when we think of bread.

So how do we navigate this as consumers? I try to make my bread myself, but I am lucky enough to work from home, which gives me more flexibility to do this. While I realise that many people won't make the time to bake bread, I can at least encourage you to give it a try if it's something you've not done before. It's a fabulous activity to do with children and it makes the house smell divine. You can form loaves that are appropriate to your family size, reducing waste. I usually make two loaves at once and pop one in the freezer for later.

I do sometimes get caught breadless, however. If you're lucky enough to have a local bakery near you, take advantage and get to know your baker! We don't have a good local bakery that's convenient to us, unfortunately, so I tend to go for the baked-off loaves at my nearest supermarket. It doesn't concern me that the loaf has previously been frozen as I don't feel it compromises the flavour or quality of the loaf that much. However, I need to have a discussion with them about what additives these loaves include. I also always buy loaves that are at least 50 per cent wholegrain for the nutritional value and I never buy pre-sliced because additives are definitely required to keep the crumb soft after it's been sliced. As far as tortillas go, I admittedly leave that up to the food industry, as mine tend to make better Frisbees than wraps.

CHAPTER FOUR

Ripe for the Picking

I walked into a supermarket today at lunchtime to pick up a few items. It was one of the rare occasions when I didn't have my son with me and I was, quite frankly, looking for a little writing inspiration, so I took my time. Like most supermarkets, my local Tesco has placed the fruits, vegetables and fresh cut flowers near the front door. The rationale, as I understand it, is that if shoppers are greeted by a vibrant colourful display as they walk in, their mood is instantly transformed in such a way that they will want to stay and spend money ... and it is the aroma of the bakery at the back that lures them deeper into the store.

I moved slowly through the aisles of the produce section and found something missing. I wasn't really getting the sweet smell of strawberries or the earthy aromas of mushrooms. I didn't even catch the faintest hint of fresh basil as I brushed by the herbs. Sure, there was a feast of colour, but I was in an aromatic wasteland.

The reason for this, of course, is that the majority of this colourful produce is packaged up in plastic. Shrink-wrapped broccoli, bagged bananas, trimmed beans on plastic or styrofoam trays and berries in plastic punnets. It is a sea of largely non-recyclable plastic. And yet, when asked, consumers say sensory information – smell, look and feel – is what they use most to determine the quality and freshness of fruits and vegetables. For consumers, the ideal fruits and vegetables are unwrapped, not just because we can hold them and smell them, but because we can select the exact quantity we want. So why are we being denied this by these plastic force fields?

For starters, it can be considered convenient for the consumer as well as the supermarket. Morrison's, for example, wraps up quantities of produce that equal £1 so that shoppers can budget as they shop. A bag of mixed cut lettuce is more convenient for most customers than buying several whole heads of different types of lettuce, especially if the bag contains a small packet of salad dressing as well.

Wrapping some produce up also keeps premium items separate. For example, organic items in the UK often have more packaging than non-organic items, in order to make sure they aren't coded in at the cheaper non-organic price (by accident or on purpose). Could you imagine the chaos if some loose organic carrots were beside some loose non-organic carrots?!

Wrapping can also help provide more information about the product – where it was grown, barcodes, batch numbers, weight and other things.

The majority of this plastic wrapping, however, is intended to protect the product and extend its shelf life. The Waste and Resources Action Programme (WRAP) estimates that extending a product's shelf life (its 'use by' or 'best before' date) by just a single day keeps that food out of the bin and could help save as much as 250,000 tonnes of food waste. That's rather significant. Packaging can provide physical protection: a plastic punnet is designed to stop soft berries from getting squished while still letting them breathe, for example. It can create a physical barrier to contamination by microbes, or it enables the environment within the packaging to be controlled: carbon dioxide and moisture levels in bagged salad, for example. Packaging can also prevent dehydration. Bananas are all shipped in plastic bags in order to prevent them from dehydrating en route. Many retailers remove those plastic bags before putting them on the shelf, but the bananas can last up to two weeks longer if they are left in their bags.

While there is sound reasoning for packaging, there is also evidence to suggest naked fruits and vegetables are better. We get great pleasure from holding and smelling produce and use this information to judge its quality. A study from the Netherlands in 2016 even showed that people are more likely to buy produce if it's unpackaged. This is particularly true of organic produce. People most likely to buy organic produce are guided by environmental ethics. When given the choice of British unwrapped carrots versus British organic carrots wrapped in plastic, they may choose to avoid the plastic wrapping because they think that the environmental gains of buying organic are countered by the plastic packaging. We are essentially putting organics at a disadvantage on the supermarket shelf when we wrap them. But is the perception that plastic-wrapped food is evil correct?

I am fortunate enough to buy the majority of my produce from an organic producer who delivers my fruits and vegetables to my door in a reusable box weekly. Most of the items are loose in the box, but some are still wrapped in plastic. This surprised me initially and so I had to investigate. To understand this rationale, it is necessary to look at the whole lifecycle of the food – a very complex area of research, which I'm admittedly going to oversimplify.

It's agreed that there is a considerable environmental impact associated with growing food, from water and energy use to greenhouse gas emissions. An environmental investment has been made in growing that food and so it is obviously desirable to protect that investment from going in the bin or compost. Like it or not, packaging helps protect this investment. So, while we might want less packaging or more easily recyclable packaging, this might actually result in greater food waste, negating any environmental gains made on the packaging front. In fact, if packaging can help reduce the loss of food by 2 per cent, then it is environmentally beneficial, even if that packaging is 20 per cent more energetically expensive to produce.

So the odd bagged item in my organic veggie box is packaged that way to protect the environmental investment that has been made to produce it and bring it to me. I therefore have a tremendous responsibility to make sure that it does not go to waste. Those are, after all, precious nutrients.

We seem to have a love-hate relationship with vegetables. Fruit is an easier sell because it's so sweet. There have been waves in history where vegetables have gained in popularity, whether by necessity or by fashion. At the beginning of the twentieth century, the suffragette movement brought vegetables into vogue. Meat was considered masculine and so a vegetarian diet became oddly aligned with the battle for women's right to vote. During the First World War, fish and meat was hard to come by and so vegetarian versions of popular dishes, such as kedgeree, were served up. In the 1930s, everyone in the UK was detoxifying from a decade of heavy boozing (while Americans were just coming out of a decade of prohibition) and so there was a push by the UK government to promote the importance of a balanced diet; vegetables and fruit, usually suspended within moulded gelatine (Jell-o salad) became the popular food of the decade. With rationing in the 1940s, vegetables maintained their place of prominence at the dinner table in order to fill the voids left by limited dairy and meat supplies. However, once rationing was lifted, vegetable purchases started to dwindle. They dwindled through the 1950s and bottomed out in the 1960s with the revolution of ready meals. Vegetables and fruit became only slightly more popular as people dipped them in cheese or chocolate fondues in the 1970s and made wacky contrasting creations, such as cheese and pineapple sticks, in the 1980s. In the 1990s, with Bovine Spongiform Encephalopathy (BSE, also known as mad cow disease), people began to question the provenance of their food and the public grew more concerned about animal welfare, the environment and nutrition. The 'slow food' movement began. Vegetable and fruit consumption were once again on the rise, and food science had delivered a greater diversity to choose from. Supermarkets now carried about 400 different produce items, more than quadruple what could be found in a 1970s supermarket. And yet, people still weren't consuming enough vegetables.

In 2002, the World Health Organization (WHO) began to develop a global strategy that used diet and lifestyle to help tackle chronic diseases. Essentially, not eating enough fruit and vegetables was taking its toll (and still is). As a result, governments around the world began to set dietary goals around how much we should be eating. In the UK, it's '5 servings a day' of 80g each, which is equivalent to the WHO recommended 400g (14oz). In Australia, it's '2&5' of fruit and vegetables, respectively. In Canada, the recommendation is 3.5 cups of fruit and vegetables daily, while in the US it's between 2 and 3 cups of vegetables. Countries set their guidelines based on what they think are realistic targets – too high and people won't even try. A US study between 2007 and 2010 found that 87 per cent of people weren't meeting the vegetable target. In Europe, only 6 of the 28 EU states are meeting or exceeding the WHO recommendations. A 2014 study found that 12 per cent of men aged 16 to 24 and 10 per cent of women in the same age range were eating no fruits or vegetables on a daily basis; 15 per cent of boys and 7 per cent of girls aged 13 to 15 were also eating no fruits or vegetables. None. A Canadian study published in 2017 went one step further and estimated what the economic burden of not eating sufficient greens was. They found that 80 per cent of women and 89 per cent of men were not eating sufficient fruit and vegetables, which they estimated was costing the Canadian healthcare system CA$3.3 billion (£2 billion) each year. They estimated that if everyone added one serving of fruit and vegetables to their daily intake, Canadian taxpayers would save CA$9.2 billion (£5.6 billion). Get on that, will you, Justin?!* As a whole, we resist eating our vegetables, but we can be tricked into eating more.

---

* Justin Trudeau, currently Canadian prime minister (2017).

The reinvention of the carrot

In keeping with the other chapters, I should really now go into a lengthy discussion of when we first started to cultivate our favourite fruits and vegetables, how this cultivation shaped our evolution as humans and perhaps even how different methods of processing and preserving these fruits and vegetables evolved around the world. I might, for example, discuss how it was 7,000 years ago that our ancestors squished the first grapes into wine and 6,500 years ago that they pressed the first seeds for oil. It was more than 4,000 years ago that someone decided that putting vegetables in vinegar and salt was a good idea. And it was in AD 965 that someone first pressed soybeans into tofu. I would tell you about a mild version of kimchi that Koreans were fermenting more than 1,500 years ago and how it was probably brought to Germany by travellers, where it became known as sauerkraut. But instead of writing about all of that, I am going to tell you about carrots.

I personally love carrots. They are delicious cooked or raw. They are cheap, yet versatile. They can be sweet as anything, yet they are sturdy enough to get tumbled about in a school lunchbox or thrown in a rucksack while hiking. However, not everyone shares my viewpoint.

For two decades, between 1960 and 1980, carrot production in the US hovered at around 815,000 tonnes annually (give or take a few thousand tonnes). In this same period, the US population increased by over 40 million. Carrot consumption wasn't just stagnant; people were eating fewer carrots.

At the same time, some carrot farmers, well, one Californian farmer in particular, Mike Yurosek, was tired of seeing more than half of the carrots he grew go to waste. Every farmer expects to lose some of his or her crop to pests and disease and elemental factors that are beyond their control. However, what annoyed Mike no end was that he was wasting carrots because they weren't pretty enough for the stores and that was depressing. After putting in all those resources and all that time to help nurture something from seed and to watch it grow and reach maturity despite pests and disease and other hardships, he saw it rejected because it had a curve or two 'tails'. Anyone who has ever grown a vegetable in their garden can sympathise. It would be like being refused your diploma on graduation day because you had a pimple! For Mike, about 362 tonnes of his pimpled carrots weren't getting their diploma daily. He fed as many to his pigs as he could, but apparently they ate so many carrots they were starting to turn orange!*

In 1986, a light bulb went on in his head. Mike sold some of his carrots to processors who cut them up into coins and cubes for bags of frozen mixed vegetables. He then picked up a potato peeler and began whittling down his deformed carrots into little two-bite portions. It was laborious, but he had a snack-sized carrot that was pretty. He bought an industrial green-bean cutter from a local processing plant that was closing. The machine cut his ugly carrots into two-inch pieces, which meant that at least a portion of the ugliest carrots could be used. Then he shipped his two-inch carrot pieces off to an industrial potato peeler that not only peeled them, but rounded them out into uniform shapes. Mike bagged up the 'baby' carrots (not to be confused with real baby carrots, which are just young carrots) and sent them to one of his best retail customers to try out. They called him the next day and said that from now on, they only wanted the 'baby' carrots.

This was not the first time minimally processed fruits and vegetables had been introduced to supermarket

---

* While this sounds somewhat far-fetched, the carotenoid pigments that make carrots orange are fat soluble and excess amounts are stored in body fat. It was the pigs' fat that was turning orange as a result of eating so many carrots, which I suppose could be a bit off-putting. However, I'm sure that given the right marketing strategy, Mike could have started a new food fad. Just imagine, have your pork scratchings and carrots too!

shelves. It had been tried back in the 1940s, but more as a means of using up produce. Processing methods were not as advanced and so the shelf life of these cut-up and pre-peeled fruits and vegetables was extremely short. The concept didn't take off. It wasn't until the 1970s when fast-food chains began to bring in pre-shredded lettuce and chopped onions for their burgers that there became a large enough market for processed produce to make it worthwhile. In the 1980s salad bars were popular and restaurants didn't want the additional labour of preparing all those vegetables. Reducing labour meant either buying specialist equipment for slicing and dicing or outsourcing it to processors who already had the equipment. By outsourcing, restaurants saved staff time as well as organic waste. It also meant that less weight was being shipped around. It was a no-brainer.

Mike's baby carrots proved the 1980s consumer was ready to see this level of processing on supermarket shelves. Customers went crazy for the no-peel, no-fuss, bite-sized healthy snacks. It's estimated that US consumption of carrots increased by 33 per cent thanks to these new little morsels, and by 1999 more people were buying baby peeled carrots than regular carrots. Mike was happy; his minimal investment in equipment was earning him five times as much from retailers for his baby carrots. Retailers were also happy because they could sell what they eventually called 'mini' carrots for double what they paid for them; they could make 50 cents from the mini carrots whereas they were making only 7 cents from the same quantity of whole carrots. The carrot industry as a whole was happy; Mike didn't patent his product so everyone started to produce their own version of the mini carrot. Specialised equipment was made to produce the popular snack. Ten years after Mike had first peeled those ugly carrots, carrot production in the US had more than doubled. And I can only assume that Mike's pigs were also happy because they didn't have to eat quite so many carrots.

As well as reviving the inner-bunny of the American public, the mini carrots, neatly wrapped up in a plastic bag, put a brand name front and centre on carrots. Producers therefore wanted to ensure that the product behind their brand was the best, while still being efficient. This meant changing the shape of carrots. Literally. A breeding frenzy began in order to create the ultimate 'rough' carrot from which as many uniform two-inch carrots could be sculpted as possible. This new breed of carrot would be long with a stubby rather than tapered end; it would have no core, and it would be vibrantly and uniformly orange and sweet, without losing that distinctly carrot taste. Varieties of carrot that had been discarded long ago for being too long to fit into bags were revived and crossed with other varieties to produce carrots that could be cut into as many two-inch pieces as possible. They planted the carrots closer together to encourage a skinnier carrot while also improving yields for each field. The trait priorities had shifted 180 degrees; carrots that weren't skinny enough or long enough to continue on the processing line to become mini carrots were rejected and sold as whole carrots.

The new varieties of carrot were named Prime Cut, Sweet Cut and Morecuts in recognition of their primary purpose. Changing the shape of the carrot towards a long cylinder meant that less was wasted in the processing. Pieces of carrot were discarded for imperfections rather than entire carrots. The mechanical peeling process is designed to waste very little of the carrot – less than 1 per cent – and these peelings are used for animal feed. They are not chemically peeled as is often thought. Like most vegetables, however, mini carrots are usually treated with a chlorine solution to kill off harmful bacteria that can cause them to spoil. They are then washed again in water before being bagged.

The mini carrot is an example of how making a vegetable slightly easier to eat can encourage healthier snacking. It was the beginning of a flood of minimally processed products, known as fresh–cut fruit and vegetables, into supermarkets everywhere. It shifted the waste and effort associated with preparing fruit and vegetables out of households and into processing plants. And while the more highly processed fruits and vegetables – juices, purées, fermented and canned goods – are an important aspect of our nutrition, it is this minimally processed produce that I think makes for the more interesting story here because it is not necessarily as well known.

The fresh–cut time bomb

This is my experience every June when the first strawberries of the season start making an appearance. I am overjoyed to see them on the shelf after a winter of citrus and last season's apples. I examine all of the punnets in the store, staring through the clear plastic to determine which bunch is the most ripe and shows no signs of rotting. I give them a place of privilege in my shopping basket so that they don't come to any harm by touching a rogue avocado or, heaven forbid, canned goods. I purchase my carefully selected strawberries and put them in the top of the shopping bag so that there is no fear of mishaps on the 15-minute walk home. Once in the kitchen, I pull the strawberries off the top of the bag and as I place them on the counter, I notice one or two have already passed their prime and are in a state of decay. Sigh. This is what has become known at my house as the strawberry time-bomb.

It is as though I am in some culinary version of the crime thriller *24*, where every time I enter my kitchen I can see a clock counting down to the moment when my strawberries are going to implode, and I am agent Jack Bauer trying my best to defuse the situation, yelling at all family members within earshot to 'eat the strawberries'. I have even been known to lash out at people caught eating other fruits! I exaggerate ... but only slightly.

Strawberries are not alone in having a brief window in which they are at their peak quality. This is when a fruit's colour and smell sends out signals to everyone that it is time to be eaten. The fruit's sugar content is usually at its highest, luring animals in to take advantage of this energy reward in exchange for some seed dispersal. This is how birds know to eat all of the cherries or blackcurrants or other soft fruit you have nurtured in your garden on the very morning you intend to pick it. This window obviously varies depending on the fruit or vegetable in question. The goal is to deliver them to consumers as close to this window as possible, having processed the fruit and vegetables and got them to consumers so that they can eat them without going all Jack Bauer on their families. This is a significant challenge.

Some plants, including tomatoes, avocados, bananas, mangoes, plums and apples, start to produce large quantities of the gas ethylene at different stages of their lifecycle. These are known as climacteric plants. Ethylene acts like a general chemical signal within the plant – much like a hormone in animals – which triggers changes, including the opening of flowers or shedding of leaves, and, important in this context, the ripening of the fruit. Climacteric fruits reach full size before ripening; this growth is known as maturing. As they mature, the fruits store starch. Once mature, ethylene production increases and the fruits start to ripen – a tomato goes from green to red, for example. The fruits begin to respire rapidly, which triggers changes in colour, and the starch is converted into sugar. This is when this fruit is at its peak; everything after this respiratory burst is a rapid decline into rot. Climacteric fruits respond strongly to ethylene and will therefore continue to ripen after they've been picked from the plant. As you probably already know, producers can use this trait to their advantage. Bananas, for example, are picked hard and green, long before they are ripe. This makes them easier to transport. They are kept in cool storage until they are needed and then they are put in a room where ethylene is pumped in, which will ripen them up ready for retail. At home, climacteric fruits can be stored on the counter at room temperature until they are fully ripe and then placed in the refrigerator to help slow their degeneration down.

Strawberries, however, along with citrus fruits, aubergines, leafy vegetables, bell peppers, cucumbers, pineapples and grapes, are non-climacteric fruits – they do not continue to ripen after they've been picked. And this is why we have a strawberry time bomb. Non-climacteric fruits need to be picked when they are fully ripe and from that moment, the race is on!

Strawberries grown in California, for example, are picked by hand and placed directly into plastic punnets. The minute a strawberry is picked it begins to respire more,[*] using up sugar and losing moisture in the process. The picker's strawberries are inspected for quality in the field and then driven by truck to a cold-storage facility on the farm. If that truck gets stopped for any reason, it will lose 10 per cent of its cargo for every hour it is delayed. Tick-tock, tick-tock. It's not just the hot strawberry fields that cause this loss; fruits and vegetables can generate considerable heat of their own accord when stockpiled (not that you would ever stockpile strawberries unless you were intending to make jam, but I wanted you to know this great fact). At room temperature, a tonne of vegetables can generate 127,000 kilojoules of heat daily, which is equivalent to the amount of radiation one square metre of the Earth receives in 25 hours, or enough heat to melt over 360kg of

---

[*] All plant cells respire, just like animal cells. They take in oxygen and expel carbon dioxide as part of the process of releasing energy from glucose. This is as opposed to photosynthesis where plants use energy from the sun to convert carbon dioxide and water into glucose and oxygen. It's important to note that cellular respiration is not the same as breathing; as much fun as it might be to think of strawberries as heavy breathers, they are really just rapid respirers.

ice cream in a single day (which is an odd unit of measure, but I'm working with what I've been given). The point is that they generate a lot of heat and heat is no friend to the deteriorating plant. A strawberry lasts only a matter of hours in the heat, but keeping it at 2°C will extend its shelf life to about 10 days.

In the on-site cold-storage facility the strawberries are cooled using large forced-air coolers. These machines take three hours to cool the strawberries to 2°C, which would take three days in a normal refrigeration unit. The lower temperature slows the activity of the enzymes, delaying decay. Once they are cool, they are sealed in another layer of plastic. A good portion of the oxygen in the container is sucked out and carbon dioxide is added. This slows their respiration right down, as well as the consumption of sugars and water – they are in a sort of strawberry hibernation.

Three and a half hours after the strawberries are plucked from the plant on a Californian farm, they are loaded onto refrigerated transport trucks. Inflated plastic pillows are used to cushion any movement between the pallets on the truck. They travel out across the US and Canada, some travelling more than 4,800km (2,982 miles). Tick-tock, tick-tock.

They arrive at a general warehouse at their destination as much as two days and a couple thousand US dollars-worth of fuel later. They are stripped of the plastic wrap that has kept the mix of gases in their container different from the outside atmosphere and with this new influx of oxygen, their respiration rate starts to increase again. Their temperature is taken to ensure they have been kept cool as the distributor needs to know exactly how much time they have to get them out to retailers. Tick-tock. They are kept in cold storage until they can be shipped out. Once they hit the retailer they are plunged into a surrounding temperature of 20°C (earlier if they were transported in a non-refrigerated truck) where rapid respiration begins again and any fungi or bacteria present on the fruit start to flourish. It's a miracle any make it home to the consumer in one piece, quite frankly.

You will recall from Chapter 3 that I live not far from the village of Cheddar. I mention this lovely village again here because the Cheddar Valley Railway Line used to connect a number of villages in the Cheddar valley, which famously grew delicious strawberries. This railway line became known as the Strawberry Line due to the quantities of strawberries that were transported from these communities in the nineteenth century to the junction at Yatton, where they then made their way to London markets, and most importantly, the Wimbledon tennis tournament. The railway line is gone, but this route is still known as the Strawberry Line and is a favourite off-road cycling route of mine. Even then, they knew that they were working against the strawberry time-bomb.

Strawberries are clearly a bit of an extreme example with their soft fleshy skin and tendency to rot if you look at them wrong. Other fruits and vegetables have significantly longer shelf lives, relieving the pressure considerably – potatoes and apples, for example, are stored for six months or more before they reach retailer shelves. However, when you start to process them – even the simplest of chops and the lightest of peels – this introduces a whole new set of challenges in terms of keeping things fresh.

The moment that fresh fruits and vegetables are cut or peeled, their living tissues have been wounded: the skin that is there to protect them from microbial invasion has been compromised. Their cells are split open and their contents are strewn upon the surface of the cut. Within those cells are enzymes, which were nicely contained in little compartments within the cell wall, but are now exposed and begin to cause chemical reactions in adjacent intact cells. Signals are also sent out, similar to when we receive a wound, which trigger a series of responses. (Makes peeling a carrot seem far less innocent, doesn't it?)

These are some of the things that are happening at the site of the cut:

- Enzymes released by the cut cells start to degrade the cell walls of nearby intact cells, causing the flesh around the wound to soften (a bruised banana, for example).
- Other enzymes start to degrade the chlorophyll, causing greens to lose their colour (at the stem where spinach is cut, for example).
- Exposure to oxygen in the air can cause antioxidant compounds (phenols) to accumulate in the wound area. The enzyme polyphenol oxidase can then convert these into other compounds that cause the browning at cut surfaces.
- Ethylene is also produced in response to the wound, speeding up enzymatic activity and therefore degradation in the area.
- The plant begins to respire more, using up water, sugar and essential vitamins, leading to limp, flavourless produce with less nutritional value.
- And if that wasn't enough, all of this increased metabolic activity can lead to the development of other compounds, which in turn have different effects within the plant; the metabolite lignin, for example, is produced when asparagus is cut and it is this compound that toughens up the stalk near where it is cut.

As a result of the plant's response to injury, fresh-cut fruits and vegetables deteriorate faster than their whole counterparts; a whole pineapple or cantaloupe can easily last a week if refrigerated, but these same fruits cut up in the refrigerator don't stand a chance of lasting a week. By making fresh produce easier to eat for consumers, the industry speeds up the clock on the time-bomb.

Delaying the decay
Many of us have squeezed some lemon juice over a fruit salad or freshly cut apple in order to prevent it from going brown; and we may know that cut carrots are best kept submersed in cold water in order to prevent them from drying out and oxidising. These are techniques we all have for preventing our hard work from becoming unpalatable. The fresh-cut industry has far more tools at its disposal.

The first step is to move the produce from the field through to processing as quickly as possible. You will recall from the section above that stockpiling fruits and vegetables can melt a lot of ice cream, so there is as little delay as possible.

Once it is in the processing area, the produce is all washed. This isn't always done with whole items because washing can actually create a perfect environment for fungus. But because fresh-cut produce is going to be further processed, it's thoroughly washed to ensure no contaminants, such as dirt, pesticide residues, micro-organisms or enzyme residues remain on the surface. A processing facility that is simply washing whole carrots might use half the amount of water during this step compared with a facility that is peeling and cutting the carrots – it takes 1.5–3 cubic metres of water per US ton to wash whole carrots, as opposed to 3.5–5 cubic metres to wash carrots for processing. While many facilities try to recycle as much water as possible during the wash, it needs to be disinfected before being reused in order to remove any human pathogens that may have contaminated the water system. Chlorine is often used to kill microbes, but the legal level set for its use doesn't kill all the microbes. Many facilities now use ozone as an antimicrobial in recycled water systems as it dissipates quickly out of the system, but a washing-tank sanitiser is still needed to keep the equipment clean. There are pros and cons to all methods and it is an active area of research as people try to find ways to maintain high standards of food safety while reducing water consumption at this stage.

Temperature is also strictly controlled. Washing water is often as cool as 0°C in order to slow down the biochemical reactions happening within the living cells. The formation of lignin in asparagus that I mentioned above, for example, is slowed through cooling the cut asparagus spears. In other cases, low heat might be used. Mild heating of whole apples helps to keep them firm after they are cut. However, over-chilling and over-heating bring on a new set of problems, so it is a delicate balance.

The produce then goes through one or more steps – mechanical peeling, slicing and dicing, for example – to alter its appearance. This is the 'fresh-cut' part. Delaying decay at this stage means keeping equipment clean to avoid contamination and keeping knives sharp; bacteria, such as *E. coli*, seem to be better able to penetrate into the flesh of produce that has been cut by a dull knife, presumably because there is more surface area to a rough cut.

There are then a series of chemical post-cutting treatments that can be used to help delay decay and generally avoid discolouration. The brown surface of a cut apple doesn't change the nutritional quality of the apple or even its taste really in the short term, but it doesn't scream 'fresh' when you look at it either. Lemon juice works to stop browning because it's acidic and it lowers the pH on the cut surface, rendering the browning enzyme, polyphenol oxidase, useless. The food industry has its own version of freshly squeezed lemon juice that it uses to stop discolouration. Cut fruit and vegetables can be dipped in mild solutions of a variety of food-grade acids, which are made from an isolated powder, but the compound at work is still the same: citric acid. Ascorbic acid (vitamin C), acetic acid (vinegar), phosphoric acid (a mineral acid) and tartaric acid (from grapes) may also be used, or, more likely, combinations of these and others. Yet, as we know from lemon juice on an apple slice, some of these acids can leave flavours behind, so they may not always be appropriate.

Ascorbic acid prevents browning through another mechanism as well as just lowering the pH. It acts as a reducing agent, losing an electron to the browning compounds, which converts them back into phenolic compounds. When it does this, however, the ascorbic acid is used up in the process, so once it's gone, browning begins again.

There's a whole host of other food additives that can be used to stop browning in a variety of ways. Calcium disodium ethylene diamine tetra-acetate (more easily known as EDTA) attaches itself to metal ions, which prevents them from catalysing the reactions that cause browning. Salts of calcium, zinc and sodium have all been investigated for their ability to inhibit enzymes and stop browning. Food scientists have even explored the use of other enzymes in an attempt to break down polyphenol oxidase – fighting fire with fire, or enzyme with enzyme, as the case may be.

However, it's not all about discolouration. Cut fruits and vegetables can also start to change texture, which usually has to do with either the tissues breaking down or the loss of moisture. Calcium is most commonly added to preserve texture and it can come in many forms. It might be seen on ingredients lists as calcium lactate, calcium chloride, calcium phosphate, calcium propionate or calcium gluconate. The texture of fresh-cut melon, for example, can be maintained by dipping it in a weak solution of calcium chloride – the same substance used to firm up tofu. At a molecular level, this is what's happening. Fruits and vegetables naturally contain pectin, a large, structural molecule made up of lots of different types of sugar molecules. Calcium ions naturally bind to regions of pectin, forming bridges between the molecules, which causes the pectin to form a gel. Immersing fruit in calcium chloride introduces more calcium ions, strengthening the gel and keeping the fruit firm and moist. The challenge for the industry, though, is that once the calcium ion bonds with pectin, more chloride ions are left behind and this can give the treated produce a bitter flavour.

As well as chemical dips, there are edible films and coatings, which form a more permanent barrier between the fruit or vegetable and the environment. These are used on whole and cut fruits and vegetables. At the core of these coatings are the basic macronutrients that we recognise – proteins, carbohydrates and fats – which are formed into layers of matrices that create the basic barrier. Added to these, however, are a variety of different materials with different properties that change the function of the coating. These include plasticisers, which keep the coating flexible; emulsifiers, which helps keep things mixed; and surfactants, which are usually added to help the coating stick to the product in question. Edible coatings are used in other aspects of the food industry, including meat and baked goods, for different purposes, but in terms of fruits and vegetables, these coatings are applied to the surface in order to do all the things we've already talked at length about: slow the exchange of gases (respiration), stop browning, retain moisture and preserve texture, and even provide some physical protection against injury. As well as performing these functions, the coatings have to be invisible, tasteless and safe, and they have to stick to whatever they are meant to be coating. It's a very active area of research, with more than a thousand companies in the business, making tens of billions in annual sales.

Coatings aren't new. Wax coatings have been used on whole fruits and vegetables for centuries – it is thought that the Chinese were waxing whole citrus fruits with beeswax as early as the twelfth century in order to stop them from drying out. Coating materials used today are derived from plants, such as pectin or alginates and carrageenan from seaweed, as well as from animals, such as chitosan from the shells of crabs and shrimp, and whey-proteins from milk. Chitosan extends the shelf life of mushrooms, alginate combined with calcium chloride keeps lettuce crisp, corn protein combined with oleic acid (a naturally occurring fatty acid) keeps broccoli crisp and green. Wheat protein layered with fats, such as beeswax, keeps strawberries firm and fresh, while alginate coatings have been shown to keep fresh-cut pineapple juicy.

Edible films and coatings can also be used to deliver active ingredients, such as antimicrobials, anti-browning agents or nutrients, such as additional vitamins. There are even references in the literature to including additives that can enhance texture and flavour, though I can't be sure whether this is happening beyond laboratory experiments. There is also a great deal of research going on into the incorporation of functional nano-sized materials in these coatings, but more about that in Chapter 8.

The commercially available films and coatings are closely guarded recipes with names like eatFresh™-FC, Nature Seal™, Freshseel™ and Semperfresh™. These coatings can keep a sliced apple looking fresh for 26 days, and except for the unnaturally non-browning apple slice, we don't even know they are there. These coatings, depending on how they behave, may be classified as food products, ingredients, additives, food contact substances or even food packaging. For the most part, however, they are regulated in the same way as food additives. Their classification determines labelling requirements, and understanding labelling requirements requires a degree I simply don't have. I can tell you that in Canada, only priority allergens that are contained in the coating, such as whey or wheat, need to be declared. In Europe, anything considered a food additive must be listed on the packaging with their E-number. In the US, coatings can either be regulated as additives or as substances that are 'generally recognised as safe' (GRAS); there are specified limits for how much of these can be used. In my mind, the world of coatings is a grey area, not just in terms of regulation and labelling, but in terms of ethics ... but more about that in a moment.

Instead of (or maybe in addition to) dips and coatings, it's possible to gas fruits and vegetables; compounds that block the cell receptors for ethylene are very effective at delaying decay. If you recall, ethylene acts like a plant hormone that, among other things, accelerates ripening. If the cell receptors for ethylene can be blocked, the cells won't respond to the ethylene being produced. A compound known as 1-methylcyclopropene (1-MCP) has been developed for this purpose. Experiments with sliced apple have shown that, depending on the variety, treatment with 1-MCP (a gas) slows browning and maintains texture, even after 35 days. However, there is a lot of variability to 1-MCP's effectiveness. Some varieties of apple are more sensitive to ethylene and therefore to 1-MCP. Apples are also more sensitive depending on how mature they were when they were harvested. Exactly when the apple is treated with 1-MCP also makes a difference; if the apple isn't going to be cut up immediately after harvest, then it's most effective to treat the whole fruit with 1-MCP right after harvesting, rather than three months later just before it's sliced. Research is ongoing.

Consumer distrust of additives and chemicals with long names has also pushed the industry to look more closely at more natural additives, particularly plant extracts with label-friendly names. Compounds isolated from ginger and capsicum have antimicrobial properties and pineapple juice and rhubarb juice are effective at preventing browning. It seems as though the industry might have to resort to squeezing lemons after all.

So, to quickly recap, the fruits and vegetables have been cooled (or mildly heated), thoroughly washed, sliced and diced with very sharp knives, and most likely dipped in some cocktail (natural or otherwise) to help preserve them. They've possibly been exposed to ethylene blockers and potentially coated in an edible film of sorts. Now it is time for packaging, and here the buzzword is MAP, which stands for modified atmosphere packaging.

These packages are designed to do just as the name implies: change the environment surrounding the substances they contain. When it comes to fresh-cut produce, as previously mentioned, the main goal here is to reduce respiration rates, which is done by changing the oxygen-to-carbon dioxide ratio. This can be done passively. As the produce respires, it will expel carbon dioxide into the container. If this respiration rate is known, then the permeability of the packaging can be modified so that it keeps most of this carbon dioxide in and only allows a small amount of exchange with the outside atmosphere to ensure some oxygen is still available.

This can also be done more actively by pumping different gas mixes into the package – fruits and vegetables are given a mix with a high ratio of nitrogen, for example – or incorporating materials into the packaging that scavenge oxygen. In 2013, the European Food Safety Authority approved OxyFresh technology, which are sachets, pads or labels that contain materials that are activated when they become moist and then begin to produce oxygen. The technology also absorbs carbon dioxide. The result is that the packaging converts the atmosphere within it into a high-oxygen environment, which is counter-intuitive. However, this puts the fruits and vegetables into 'oxygen shock' and suppresses ethylene production, doubling the shelf life of the produce. There are other products that scavenge oxygen or carbon dioxide, depending on what's needed. Sachets or multilayered paper or cardboard can be incorporated into the packaging to adsorb and oxidise ethylene gas given off by the produce. There are moisture control systems and pads that increase airflow around the produce as well as cushioning them. Some packaging contains materials that give off volatile substances* that

---

* Volatile sounds bad, doesn't it? In this instance it simply means a substance's tendency to go from a liquid to a gas. Many plants produce volatile organic compounds, which include aromatic compounds that we thoroughly enjoy. Rosemary, tea tree, lemon and lavender are all examples that are commonly used in fragrances because they so readily turn to vapour.

have antimicrobial properties. However, as with most of the edible coatings, the technology behind a lot of packaging is commercially sensitive. Information is given in generalities, leaving consumers quite clueless as to what is happening inside what looks like a rather inert container.

As well as keeping things fresh, the packaging has to allow consumers to see the product. This is particularly important for fresh-cut produce as consumers rely heavily on the look of the product to judge its quality. The packaging also has to provide the consumer with relevant information – 'best before' date, for example, though 'best before' dates might become redundant as the new generation of packaging comes into circulation. This 'smart' packaging has sensors built in to indicate freshness based on the atmosphere inside the packaging, and indicators that alert retailers if the food has gone outside a specified temperature range during its travels. Whether you are a fan of packaging or not, you have to marvel at the science that has gone into it.

It takes combinations of all of these methods in order to ensure that our fresh fruit and vegetables last long enough to travel across borders into cities where they are distributed to caterers, restaurants and retailers, purchased, and then, within a couple of days, eaten.

The prunes and corns of minimal processing

Fresh-cut produce is a fast-growing sector of the market. Bagged salad sales, which account for roughly half of all fresh-cut sales, grew 560 per cent in the 1990s in the US; by 2015/16 bagged salads were worth US$3.7 billion in the US. That is the kind of growth that food industry fantasies are built on. So with all of this convenient greenery going out the supermarket door, are we getting healthier? Are we actually eating more fruits and vegetables because they are conveniently cut up for us, or have we simply substituted them for the less convenient fruits and vegetables?

With the global economic burden of cardiovascular disease and cancer estimated at £664 billion (US$863 billion) and as much as £1.9 trillion (US$2.5 trillion) respectively, governments worldwide are keen to promote preventative actions that can reduce the prevalence of these diseases. Not eating enough fruits and vegetables is one of the top 10 risk factors for global mortality, and the WHO estimates that 1.7 million deaths worldwide (and the associated economic burden) could be prevented through the increased consumption of these foods. Diets high in fruit and vegetables have also been linked to lower prevalence of chronic inflammatory diseases, such as rheumatoid arthritis; metabolic disorders, such as type 2 diabetes; eye disease, such as age-related macular degeneration; and obesity. Plants contain the nutrients we hear about all the time: vitamins, minerals, macronutrients. But they also produce compounds (known as phytochemical compounds) that actively neutralise the free radicals that cause oxidative stress in the body, which is linked to degenerative diseases and ageing. So, surely the greatest benefit of minimally processed fruits and vegetables has to be making them more accessible and therefore promoting increased consumption, right?

A survey conducted in 2016 in the US found that 66 per cent of respondents said they were eating more healthily than they had been two years before, with pre-packaged salads being largely credited for offering healthier fast-food options. Yet when we look closely at the numbers, people may not be leading the healthier lifestyle they think. Bagged salad is largely comprised of leaf lettuce, as opposed to head lettuce, which is sold whole (admittedly, some is also used in salad mixes). According to the US Department of Agriculture's Economic Research Service, in 1985 – the first year they have records for leaf lettuce – head lettuce consumption was 10.7kg (23.7lb) per person annually, while leaf lettuce was 1.5kg (3.3lb) per person, a combined value of 12.2kg (27lb) of lettuce per person. In 2016, the ratios had changed considerably. Sales of head lettuce were down to 6.8kg (14.9lb) per person while leaf lettuce had quadrupled to 5.6kg (12.3lb) per person. However, combined, this comes to

12.4kg (27.2lb) per person, a negligible increase over 21 years, despite everyone saying they are eating more salads now than ever before. The same is happening here in the UK – head lettuce sales have been replaced by bagged lettuce sales.

The 2015 report *State of the Plate* looked at trends in US fruit and vegetable consumption. It turns out it is quite variable, but in general, consumption has been decreasing since 2010. If you look at specific vegetables, four different trends are apparent. There are vegetables that, no matter what, don't fluctuate in consumption levels. These are the Brussels sprouts and pumpkins that people feel obliged to eat during particular holidays, but that rarely get any attention otherwise and have not (yet) been the focus of any health-food crazes. There are a heap of vegetables that experienced significant increased consumption during the 1990s as people became more health conscious, but have since reached a plateau; these are vegetables such as asparagus, broccoli, cauliflower and spinach. There are a few vegetables that have absolutely soared in popularity since the 1970s: people are eating ten times as many mushrooms now as they did in the 1970s,* five times the number of bell peppers, one and a half times the fresh tomatoes and five times more squash. And finally, there is one vegetable that has gone completely out of favour and is singlehandedly driving the downward vegetable consumption trend: the potato. People now eat half the quantity of fresh potatoes that they did in the 1970s. Their reputation as a high glycaemic index food and enemy of low-carb diets has taken its toll.

So, despite efforts to make fresh fruit and vegetables more convenient, Americans aren't eating any more than they did 20 years ago. The UK's *National Diet and Nutrition Survey*, conducted in 2014, suggests things aren't much different here in the UK. Only 30 per cent of adults are

---

* To be fair, there were probably quite a few mushrooms being consumed during the 1970s, just not the species that are normally tracked by the USDA.

meeting the recommended fruit and vegetable intake, and among children aged 11 to 18, only 10 per cent of boys and 7 per cent of girls are meeting the '5-a-day' recommendations. This probably isn't making any noticeable impact on the prevalence of chronic diseases.

If people aren't necessarily eating more lettuce, perhaps they are eating better lettuce. Varieties of head lettuce, such as iceberg and cos (romaine), seem to be somewhat inferior in terms of certain micronutrients compared with the varieties of dark Italian leaves, rocket and watercress that are found in bagged lettuce. Perhaps some of the variety provided by minimal processing – from mixed greens to stir-fry mixes – is improving nutrition, but how is processing affecting those nutrients?

This question is somewhat like complaining about the quality of the in-flight meal when the plane is crashing; it's irrelevant if carrots have a few milligrams less of vitamin C after they've been chopped if the majority of the population isn't eating carrots at all. The first step is to get everyone eating the recommended daily intake of fruits and vegetables; after that, perhaps it's time to discuss which preparation methods deliver the highest nutritional content. However, that being said, if you're reading this book, you probably care about food and therefore you are part of the population munching these plants. So, for you, here is a little bit of information about how minimal processing affects our produce.

Minimal processing, because there's no heating or other extreme treatment of the fruits and vegetables, should mean these items have a similar nutritional value to the whole product. However, there are some exceptions. Vitamin C content seems to generally decline after a vegetable has been cut or peeled, but some vegetables are more affected than others. For example, after two days in cool storage, cut celery has about half the vitamin C content of its whole counterpart and cut carrots have one-fifth as much. Chopped red cabbage, however, has a slight increase in vitamin C content, while radishes have no significant change. It's amazing what affects this too. Using a smooth knife rather than a serrated blade results in 18 per cent more vitamin C being retained in an iceberg lettuce, and if it's cut mechanically 25 per cent more is retained than if it's cut by hand. How it's packaged also affects the nutrient content. Modified atmosphere packaging helps to retain more vitamin C in green peppers, but causes greater depletion in potatoes. The general trend, with exceptions noted, is that as soon as a fruit or vegetable is cut from the plant, its nutritional value begins to decline. At a certain point, in fact, it becomes more nutritionally advantageous to eat frozen or canned goods rather than a vegetable several weeks old that has been kept fresh with dips and coatings and fancy packages. However, this is really getting into the minutiae of nutritional quality. Eating minimally processed fruits and vegetables is most definitely better than eating no fruits and vegetables.

Lastly, are there environmental benefits to fresh-cut over whole produce? Again, this is so very dependent on a number of variables that the life cycle of each product would need to be considered. I must once again bring up carrots as they seem to be the most studied vegetable when it comes to comparing the environmental impacts of different processing methods. Farming carrots is undeniably where the biggest environmental impact is, but this will be equal regardless of whether the carrot is whole, fresh-cut, frozen or pureed, so let's remove this from the equation. For fresh carrots (whole or fresh-cut), the next biggest impact is in their transportation, including from the retailer to the home of the consumer. Two things about this surprised me in my research. First, the per kilo impact of transporting carrots is much less for a large transport truck driving a heavy carrot cargo across the country than it is for a small van driving a couple of boxes of carrots to a local farmer's market. Second, transporting peeled and prepared carrots has a lower environmental impact than transporting bunched carrots because the weight of the waste (peelings and tops) is not being transported.

As previously mentioned, the best way of protecting the environmental investment made in growing produce is to not let it go to waste. With fruits and vegetables, the second biggest loss is at the point of production (17 per cent)*. Disease, infestations, poor yields and other challenges in the field create losses rather than waste as the food was never edible. However, a significant portion of perfectly edible food is also rejected for aesthetic reasons. Another 6 per cent is lost in post-harvest handling and 9 per cent more in processing. The more the produce is handled, the more likely a machinery mishap is adding to the waste. Alternatively, though, if the food isn't handled sufficiently to prepare it for distribution, significantly more can get wasted during the next phase of its lifecycle. Distribution and retail losses account for about 7 per cent of food waste. Bagged salad in particular doesn't do well in this part of its lifecycle, largely due to overproduction. If a supermarket suddenly changes its mind about how much bagged salad it wants or the processor simply makes too much, it might not have the time to find another outlet. A supermarket chain wants to have the bagged salad in the warehouse at least 10 days before its 'best before' date so that it has time to distribute it to its various stores, get it on the shelves and give consumers enough time to buy and consume it. This means that bagged salads well within their 'best before' date are suddenly at risk of being wasted. Many are donated to food banks, but a lot also ends up in landfill. The greatest waste, however, happens with the consumer. Around 61 per cent of food waste happens in the home: bagged salads forgotten in the crisper drawer, strawberries decaying in the back of the fridge and citrus fruits shrivelling in the fruit bowl. Compare this with the 5 per cent of food waste that

---

* These are the FAO's 2011 values for North America and Oceania, which, as a region, is world-leader in food waste, producing 42 per cent of global food losses.

happens in the home in sub-Saharan Africa. It's disgraceful what households in developed countries throw away.

It's possible that providing, for example, a mixed tray of vegetables for a stir-fry helps to reduce this waste at the consumer level, but to my knowledge no research has shown this. A package that is used up in a single meal is less likely to get wasted. The UK supermarket Waitrose reduced bagged salad sizes in order to help reduce the amount of waste on that front, with considerable success. The other side of this, of course, is that if fruits and vegetables were loose, consumers could choose exactly the quantity they wanted rather than being forced to always purchase three peppers or three garlic bulbs.

There are also some waste reductions made by shifting the trimmings to the processing plant. Here, some trimmings might get redirected into other products and used for juicing, purees or, as a last resort, pet food. At home these same trimmings probably end up in the compost, which is fine if it's being put to use, but often it ends up in the landfill. I'm sure there are also some energy efficiencies to be made. Tonight, I boiled two whole artichokes on my stove top for nearly 30 minutes. We sat around and took turns peeling leaves off and dipping them in melted butter. After dinner, we had a mountain of artichoke waste, and I had a son who definitely did not think all of that fuss was worth the tiny bit of flesh he scraped off the leaf with his teeth. There's no doubt in my mind that it would be far more energy efficient to process the artichokes en masse, canning the hearts. But sometimes, the most efficient way isn't the most fun way.

I choose to get my produce from an organic supplier who delivers mostly UK seasonal products weekly. I like the fact that my carrots are dirty and there are sometimes critters nestling into my lettuce – it makes me feel closer to the farm. But I'm not going to hide the fact that it sometimes gets very stressful having to use up some of these items before they go off. After the fourth week of receiving acorn squash in winter, I start to lose it. This is not for the faint of heart. So, I appreciate that this is not going to be the right choice for a lot of people. And if someone's weekly vegetable quota is mainly coming from chips and mushy peas, then I'm glad that there are some convenient options out there that could help them to meet their '5-a-day' requirements, and potentially reduce their risk of degenerative chronic disease and the associated burden that puts on society.

In terms of understanding what the best choices are in terms of nutrition, carbon footprint, sustainability, ethics and whatever other measure you want to use to guide your purchasing decisions, it seems as if this requires considerable homework and a case-by-case approach.

My greater concern lies in the various methods that are used to slow the decay of fresh-cut produce. This, to me, is a grey area. With the best intention of increasing the shelf life of a product, an opportunity to deceive consumers has been created. In pre-packaged produce this could be resolved by having a 'cut on' date as well as a 'best before' date. This becomes more difficult when there aren't labels. If I were to order a dessert in a restaurant that had beautifully spiralled apple rings as a garnish, I would make the assumption that this apple had been cut up in the restaurant just before serving. However, it's equally possible that these rings were made in a processing plant a couple of weeks ago, coated in various things to keep them looking fresh and shipped to the restaurant. And that would be deceiving. Sometimes, I feel as though just because we *can* extend the shelf life of produce, doesn't mean we *should*. Surely there are other ways that we can reduce the amount of food that we waste.

CHAPTER FIVE

Processing Protein

Throughout human history, consuming protein has been part of a display of wealth and success. And when I say protein, I mean suckling pigs, lambs, pheasants and other creatures skewered and roasted – nobody ever put a boiled egg on a platter at a feast with pride. In the early 1900s the British diet was four square meals of protein daily, and the classic English breakfast contains at least three sources of protein to this day. In fact, perhaps the route to getting people to eat more vegetables is to make them taste like protein – bacon-flavoured spinach is surely not an impossibility. Between a flavoured coating, some aromatic compounds and a virtual reality headset, I'm sure people could be deceived into consuming their daily recommended intake of greens. I think, however, the same might be accomplished if everyone was required to prepare their food from its living form.

My sister became a lifelong vegetarian when she saw my mother plucking chickens. It was the moment she realised that the animals running about the farm were the tasty morsels of meat on her plate. She never looked back. My ethical compass is not as sturdy as hers and despite many long stints as a vegetarian, I have settled on being a conscientious omnivore (although I am aware that this might be viewed as an oxymoron).

The global consumption of meat* has been steadily rising over the last 50 years. In the mid-sixties the global

---

* A quick note here to say that strictly speaking, the definition of 'meat' is 'the flesh of an animal'. And yet fish and invertebrates are

consumption of meat was just over 24kg (52.9lb) per person per year, and in 2015, this value had risen to over 41kg (90lb). The biggest growth has been in developing countries where consumption has tripled during that time period. Globally, the consumption of beef hasn't really changed, but there has been a small increase in pork consumption and the consumption of poultry (mainly chicken) has quadrupled. Our consumption of fish has also increased, going from approximately 9.9kg (21.8lb) in the 1960s to over 20kg (44lb) per person annually in 2016. Between fish and meat, we're eating an average of 61kg (134lb) of animal each year ... each. This overall growth in protein consumption is driven by increasing demand from an expanding middle class in developing countries.

Many books have discussed the changes we have made in the animals we rear for food: how we have bred animals for particular traits, including a capacity to grow quickly, resist disease and have desirable fat-to-muscle ratios. These books raise concerns over stocking densities, husbandry techniques and the welfare of animals reared on intensive farming facilities. The issues of hormone and antibiotic use, particularly the latter in its contribution to the development of resistant super-bugs, have also been covered. I've chosen not to include these discussions in this book not because I don't think they're important, but rather because I think there's already lots of information available for people who want it. This is about what we do with the animals after they have been slaughtered.

As with any food, meat, fish and seafood (which I will collectively call protein sources) fall somewhere along a

never included in the meat category. This might explain why during my vegetarian years I was often served a fish dish as the vegetarian option. Just for the record, fish are animals. As are crabs, prawns, mussels, snails, squid ... well, you get the idea.

processing spectrum and they are all processed to some degree, otherwise Sunday roasts would be a rather bloodthirsty affair. Fish, more than any other animal, seem to occupy all the places along the processing spectrum, from the sale of whole fish to the unrecognisable surimi (imitation crabmeat being a classic example of this). However, when people refer to processed meat, they generally mean meat that has been manipulated to either extend its shelf life or change its taste. And to do that, humans have traditionally dried, smoked, cured or salted it. But before humans could do any of that, they had to catch the meat.

The birth of an omnivore

Whatever your current views on meat consumption, the general consensus within the scientific community is that we humans owe who we are today to our early ancestors who, sometime between five and two million years ago, began to chomp the occasional animal. Like our other ape relatives, humans evolved from a long line of herbivores. We know this because the alimentary canal – all the way from mouth to anus – of primates is designed for moving food through relatively slowly so that as many nutrients can be absorbed as possible. This is consistent with a high-fibre vegetarian diet.

Many of our distant primate relatives have become quite specialised in their digestion. Howler monkeys, which you'll read more about at the end of Chapter 7, have a really large colon that helps them to digest leaves. Colobus monkeys have a compartmentalised stomach that essentially makes them the cows of the canopy. Gorillas and orang-utans have made such a habit of eating vast quantities of nutrient-poor foliage that their body sizes have had to increase in order to accommodate their guts. This large body size makes them less mobile and less social than some of the other apes.

The human digestive tract, on the other hand, is not really specialised at all. We have lots of other wonderful features that make us unique among the apes – for example, a nuchal ligament that stabilises our head, long legs and a plantar arch, all of which makes us good long-distance runners. But our guts are nothing special and not particularly efficient at digesting plant materials, which is why we tend to produce quite a bit of gas (although gorillas have definitely been known to blow off from time to time). A human's large intestine accounts for around 25 per cent of the overall gut length, compared with a chimpanzee's, which is about 46 per cent. The small intestine, where 95 per cent of our nutrient absorption happens, makes up 56 per cent of our gut length, which is about twice as long as that of other apes. It means we are less capable of digesting cellulose, but we are considerably more flexible in what we eat. It's as though we are specialised at being generalists.

The early hominins probably incorporated small amounts of protein into their diet, much as modern apes do now – grubs and bugs gathered along with the vegetation and perhaps the odd bird or lizard thrown in. However, by the late Pliocene (extending from 5.33 to 2.58 million years ago), early hominins were getting more diverse in their protein consumption.

A rather unique and dense pile of bones found in the Turkana basin in northern Kenya provided evidence that at around 1.95 million years ago, *Homo erectus* was actively butchering a diversity of animals. The archaeological site exposed 740 identifiable fossil bones, about 6 per cent of which have evidence of either being cut or being pounded with stone tools, which were conveniently found in the same region and time period. The marks on the bones are consistent with crushing in order to access bone marrow, disarticulation of large animals to make pieces more manageable, skinning and gutting. The bones they found represented species from 13 taxonomic groupings: aves (bird), bovidae (antelopes and gazelles), hippopotamus, suidae (boars and pigs), equidae (horses and zebras), giraffe, rhinoceros, old world monkeys (such as macaques and baboons), elephants, bony fish including specifically an air-breathing catfish, crocodiles and turtles. There was clearly a lot more on the menu than a few grubs and termites, and the protein sources were both aquatic and terrestrial.

Homo erectus is the oldest early human to have body proportions much like our own. They had relatively longer legs and shorter arms, which were adapted for life on the ground, and they had the nuchal ridge where that critical ligament attaches to stop the head from bopping about while running. These humans were accomplished runners and this may be why they were successful at incorporating a greater diversity of meat into their diet, although it is debated as to whether they were chasing the meat down or running about and scavenging it.

These early hominins were also starting and likely controlling fires, which meant they were probably cooking some of this meat to make it easier to chew and digest. It takes one-third of the energy contained in raw meat to digest it. Cooking meat reduces the amount of energy required to digest it right from chewing through to ... well, you know. As a result, the net energy gain of cooked meat is higher. Cooking the meat also reduces the chance of food poisoning, particularly if the meat was scavenged, and killing off the bacteria means that meat can also be stored for longer. This is probably relatively important when a member of the family brings back an elephant or rhinoceros to share.

Animal flesh, bones and organs are so rich in protein, micronutrients and essential fatty acids, that *Homo erectus* had to eat far less than it would on a strictly vegetarian diet. This freed up some time and energy. *H. erectus* didn't have to push as much stuff through its digestive system in order to get the same nutrition, which meant it didn't get big like the gorillas of today and could maintain its active runner form. It is still debated as to whether the addition of meat alone, without cooking it, would have given *H. erectus* advantages that helped to distinguish it from other early hominins – adaptations for running, a bigger brain, and smaller teeth and jaws. If they were cooking the meat as well, there would have been further advantages of less chewing and quicker absorption of nutrients. Better nutrition no doubt meant their offspring were more robust and more plentiful. All of these changes contributed to *Homo erectus*'s success and, ultimately, to our success as a species.

Meat is generally a hard-earned food, however. Even if it is scavenged (a polite biological term for stealing food) there is always the risk of annoying the owner of the kill or having to fight off other scavengers. It is therefore extremely advantageous to store that meat to make sure it's accessible for as long as possible. Cooking would have helped a little, but cooking only extends the shelf life of meat for a short period of time. Freezing is very effective, but not an option in all seasons or all locations. To store meat for longer, early cultures would have had to look towards other processing methods, such as drying, salting and fermenting. And one of the earliest known processed meats was all of these methods wrapped up in one convenient package: the sausage.

The big banger theory*

Fast forward now from *H. erectus* to our own species. While there is archaeological evidence of cutting, pounding and cooking meat, we wouldn't necessarily expect to find evidence of curing or preserving; there are stone mills and clay cheese-strainers, but to my knowledge no ancient stone sausage-maker has ever been found. Fatty acid analysis of potsherds provides evidence of meat and fish being kept in clay pots, which could have been fermented, but it could also just be evidence of a tasty stew. Who's to say?

---

* If you aren't familiar with British culture, this might sound a bit racy, so just to clarify, a banger is a sausage, and indeed one half of a favourite British dish, bangers and mash (sausages and mashed potato). They are so nicknamed because if the filling isn't mixed properly and the water inside isn't bound to either the meat or some other ingredient, then it will turn to steam in cooking and cause the skin to explode ... a banger!

It is unclear exactly when people figured out that salting organs and muscle and even blood and stuffing it into intestines to smoke or dry was a good idea, but they did. Sausages would have been a good way to use up remnants and less palatable pieces of animal, making sure nothing went to waste, and also of preserving and transporting those bits. Homer's *Odyssey* includes some of the first written references to sausages as he describes smoked goat-paunches (belly) filled with blood and fat. There is also written evidence from China, from around 2,000 years ago, but it wasn't until the fifth and sixth centuries AD that recipes for making sausages were documented. During the reign of the Roman emperor Nero (AD 54–68), entire festivals revolved around sausages. In fact, the Latin word for sausage is 'botulus', after which the bacterium *Clostridium botulinum*, the cause of botulism, is named. It was named the sausage bacterium because it was thought to be the source of 'sausage poisoning', which was a common problem in the eighteenth and nineteenth centuries.

Today, most countries around the world have some form of traditional sausage, with ingredients and methods reflective of their cultures and regions – from traditional Korean blood sausages (*sundae*) that are steamed, to spicy cured Mexican sausages (chorizo), to cooked German sausages (frankfurters), which require considerable amounts of garnish to make them palatable.

The basic recipe for sausage is exactly as one might expect. There is a selection of raw materials, which will most certainly affect the outcome of the sausage depending on the animal and body parts used; the ratio of fat; any added ingredients, such as salt and spices; and of course the quality of all those ingredients. How these ingredients are then blended and how coarsely or finely they are chopped will affect the texture and taste of the sausage. The type of casing used and how densely it is packed with the sausage mixture determines the style of the final product. Once the sausage is made it can be smoked, or not. It can be inoculated with moulds, or not. It can be cooked right away and served, or it can be fermented and dried over a long period of time. How and where it is stored affects its flavour and quality and perhaps even contributes to the likelihood of its causing 'sausage poisoning'. As we have seen with cheese and bread, a few tweaks to this basic method can result in a huge diversity of foodstuffs.

There are fresh sausages, which need to be cooked before they are eaten and which have a short shelf life – German bratwurst or just your standard pork sausage are good examples. Uncooked smoked sausages – Polish *kielbasa*, for example – also need to be cooked before eating, but they keep for a few days longer than a fresh sausage as smoking extends the shelf life. Cooked smoked sausages, such as Italian bologna, don't need to be cooked before eating and they have a longer shelf life than uncooked smoked sausages. Dry sausages, such as pepperoni, don't need to be cooked or refrigerated and they last a very long time. Semi-dry sausages, such as summer sausage, don't need cooking and don't need to be refrigerated (though they will last longer if they are), but due to their moisture content they don't last as long as dry sausages. The only commonality for the range of sausage products out there is that the inside bits are ground and salted, and they are usually stuffed into some form of casing.

Compared with cheese or alcohol or even bread, there is very little information about the cultural origins of the different varieties of sausage. Homer's earliest description of a sausage sounds very similar to Scottish haggis or black pudding and probably wouldn't have been cured or dried, but cooked and served right away. The Chinese were already fermenting soya to make soy sauce by the time they were writing about sausages, so their early sausages were probably fermented, even if they didn't know what fermentation was or the diversity of organisms responsible for it. When I think of a traditional sausage, it's dry and wrinkled. There is a white mould covering the outside and it is hanging from an old nail with a bunch of other sausages.

It's most likely that the raw ingredients of the first sausages would have been salted because salt was known to help preserve meat and help with flavour. What the early sausage makers didn't realise, however, was that the salt was also encouraging the growth of lactic acid bacteria, while preventing the growth of less desirable microbes. Like the bacteria found in cheese, the bacterial cultures that occurred in these first naturally fermented sausages would have varied slightly from region to region and probably from batch to batch.

The lactic acid bacteria ferment the carbohydrates in the sausage, producing lactic acid as well as other compounds, such as ethanol and acetoin, which contribute to the flavour, as well as the colour and texture of the sausage. Meat contains very little carbohydrate – only about 1 per cent – and this is in the form of glycogen, which is converted to glucose as the muscle continues to metabolise for a while after the animal has been killed. The bacteria also break down amino acids in the protein and the fat.

Two other groups of bacteria are also found in fermented sausages: Micrococci and Staphylococci. These bacteria are thought to produce a number of different compounds that prevent fats from going rancid, as well as contributing to the flavour and aroma of the sausage.

Within the first three days of fermentation, all of these bacteria thrive, reaching densities of around 100 million per gram of sausage, at which point they tend to plateau. They are not only abundant but also diverse; studies of naturally fermented sausages from Italy (in which no starter cultures have been used) have revealed hundreds of different strains of lactic acid–producing bacteria in one sausage alone. Between the acidity and the sheer abundance of lactic acid bacteria, most other bacteria – the types that might make people ill or spoil the meat – have a hard time growing.

The other organisms important in sausage fermentation are yeasts and moulds. Like bacteria, yeasts degrade complex compounds into simpler chemicals, many of which contribute to the aroma and flavour of the sausage. They are also very important at preventing too much lactic acid from building up. While the lactic acid produced by bacteria gives a nice tang to sausage, too much can ruin the flavour. Yeasts metabolise the lactic acid, helping to neutralise the pH and develop a sweeter product.

Moulds, which require oxygen, grow only on the surface of the sausage and therefore form a protective barrier that prevents light from penetrating into the sausage and breaking down the fats through lipid oxidation, which would cause rancidity and discolouration. The mould layer also slows down moisture loss and promotes more consistent drying. So long as this mould layer forms properly, it will stick to the surface of the sausage, even when handled.* Again, like cheese, these surface moulds are very reflective of the region where the sausage is made. Traditionally, the source of these moulds would have been in the ripening rooms where the sausages were hung to dry.

As with all of our food, we have tinkered with the traditional method of sausage-making in order to produce more sausages for less money in a shorter period of time. The plus side of converting this craft into an exact science is that 'sausage poisoning' is largely a thing of the past, and a better understanding of the microbes at work and the reactions happening within the sausage milieu makes it easier to control the process in order to get the best outcome with the least waste. Commercial sausage manufacturers also have to navigate food safety standards and meat regulations, as well as modern consumer tastes; for most of us, salted organs in intestine aren't particularly appealing. No longer is the humble sausage a means of using up the less desirable (though no less nutritious) bits and bobs.

---

* The particular reference I read for this information suggested that this was true so long as one didn't 'abuse' the sausage during storage. So, please don't abuse your sausage. You've been warned.

There are certainly more rules now about what can go into a sausage than in Homer's time. If sausages are labelled as pork, for example, the USDA states that they can't contain pork by-products (non-muscle meat). In 'breakfast sausages', however, animal by-products are fair game. American blood sausages can contain blood and snouts and lips, but if they don't contain any meat then they need to be labelled as blood pudding. The UK continues to embrace the original function of the sausage, and liver, skin and tongue often make their way into sausages. Non-meat ingredients must be specifically declared on the label, and not just generically labelled as offal, but rather as what they are, for example 'beef liver'. The non-meat ingredients also can't count towards the minimum meat content requirements, preventing sausage makers from filling a premium 97 per cent pork sausage with mostly liver. There are limits to everything though. Even in the UK there are body parts that are never allowed in uncooked meat products, including brains, rectums and testicles. This puts a lot of urban myths to rest (and perhaps starts a few more).

Regulations also stipulate the minimum amount of meat required, the ratio of connective tissue to meat, and the maximum percentage of fat that is allowed. In the US, the amount of water in sausages is also regulated. Anything more than 3 per cent must be declared on the label as 'added water', though this might vary depending on the style of sausage. In the UK, this isn't necessary for a meat product like sausages, but it is required on whole cuts, such as a chicken breast, where consumers wouldn't expect water to be a normal ingredient.

For fermented sausages, starter cultures are now added to provide the ideal microbial community for achieving the desired flavour profile in the desired time. To speed up fermentation and therefore the production of lactic acid, sugar is added to sausages to provide additional carbohydrates for the bacteria to ferment; sugar also helps to counter the hardening of the meat caused by salt. Most sausages these days will have a couple of different types of sugar listed in the ingredients, even if they aren't fermented.

And sugar isn't the only new addition. As well as salt, nitrates or nitrites might be added to cure the meat. If nitrate is used, it is reduced to nitrite by bacteria in the meat – something Micrococci and Staphylococcus are particularly good at. Nitrite reacts with the protein myoglobin, which is found in the muscle tissue and, like haemoglobin, binds to both iron and oxygen; it is responsible for the colour of red meat. Nitrite binds to the iron atom in the myoglobin, preventing it from losing an electron, which is what happens when fresh meat turns brown; nitrites help to stabilise or fix the colour in meat products. As well as doing all of this, nitrites are an antibacterial, they slow rancidity and they develop the characteristic flavour of cured meats. The major drawback is that several adverse health problems have been linked to nitrite exposure, particularly a higher risk of cancer. The carcinogen nitrosamine can form when nitrite reacts with amine compounds that are naturally formed when proteins degrade. This is more likely to happen at higher temperatures, which is why cured meats that will be cooked at higher temperatures have very restricted nitrite levels.

The safety of nitrates and nitrites flares up periodically in the media, where headlines equate eating a couple of rashers of bacon each morning to smoking a couple of cigarettes. This is definitely blowing the issue out of proportion. All vegetables naturally contain nitrates and nitrites. Radishes have relatively high nitrate concentrations, as do butterhead lettuces, while beetroots are high in both nitrates and nitrites. Nitrate has been linked to reduced blood pressure and potentially to a protective effect on blood vessels, such as anticoagulation. For this reason there are a plethora of beetroot products on the market targeted at endurance athletes to help improve performance. In the meat world, manufacturers are turning to vegetable extracts, such as radish or cabbage juice, in order to develop that same cured meat colour and flavour. Nitrates are still the cause, but it looks a lot cleaner on the label to say 'no nitrate or nitrite added except that found in radish juices'.

As well as nitrates and nitrites, other preservatives and antioxidants are used to extend the shelf life. Propyl gallate (E310) is a synthesised additive that prevents the oxidation of fat, for example. Butylated hydroxyanisole (BHA) is a powerful antioxidant sometimes added to sausages for the same reason – it's really good at protecting fat from free radicals. Sodium erythorbate (E316) is also frequently used, particularly in items like hot dogs, because it helps prevent the formation of nitrosamines by facilitating the conversion of nitrite to nitric oxide.

Among the other additives one might find in today's sausage are: phosphates, such as hexametaphosphate (HMP) or diphosphates; alkalis, such as sodium bicarbonate; and acids, including citric or acetic acid. All of these additives have a similar function in that they enhance the texture of the sausage but mainly they help water bind to protein. This keeps more moisture in the sausage and less in the pan during cooking. High phosphates have also been linked to health issues. While they are naturally occurring in all high-protein foods, adding them to things like sausages means we are ingesting them in higher quantities than we would naturally. In fact, research from Germany suggests that the consumption of phosphate-containing food additives (there are over 300 of them) has doubled since the 1990s. Higher phosphate intakes have been linked to higher mortality among patients with chronic kidney problems and with cardiovascular disease in otherwise healthy people.

To keep sausages looking and tasting nice, there are more additives. Colouring additives, so long as they don't exceed specific quantities, can include synthetic dyes, such as red 2G (E128) or naturally occurring pigments, such as canthaxanthin (E161g). Spices are added to provide flavour, obviously, but some serve other functions as well. Rosemary, cloves and cinnamon, for example, have antimicrobial and antioxidant compounds; the latter help to stop fat in the sausage from going rancid.

There are a number of different fillers (also known as extenders) that are used to reduce the cost of more expensive ingredients, but also help to bind the ingredients together and soak up any water that isn't bound by the meat. Breadcrumbs and rusk (essentially dry unleavened bread) are popular fillers. Non-meat proteins, such as soy protein or whey protein, can also serve these purposes as fillers. As they get denatured in the cooking process they interact with the surrounding meat proteins, help with water retention and reinforce gelling to improve the texture of the sausage. There are guidelines as to how much of these isolated proteins can be added.

Many sausages also contain various polysaccharides, long-chain carbohydrates that come from plants. Examples include carrageenan and alginate (both from seaweed), locust bean gum (from the seeds of the carob tree), xanthan gum (fermented by bacteria) and starch (usually potato). These large molecules are particularly good at binding water and encapsulating fat particles, so they essentially give low-fat products a much more fatty mouth-feel. This makes them a particularly popular addition to low-fat sausages.

It's not just the inside of sausages that have changed. Natural casings are still made from the intestinal tracts of farmed animals; they've had the outer layer of fat and the inner lining removed, and they are washed in water and salted, but otherwise that's pretty much the extent of the processing. They haven't changed for hundreds of years, although there are now machines that do a lot of the dirty work. However, some people get a bit squeamish about intestines and in response to this a whole bunch of artificial casings have been developed.

Edible collagen casings were the first to go into commercial production nearly 60 years ago. Collagen (a structural protein) is extracted from waste material such as specific layers of the skin, bones and tendons of animals. Through some fancy chemical processing collagen is formed into a dough that is not dissimilar from gluten. This is then extruded into intestine-like lengths of different diameters for filling. Collagen casings are much cheaper and quicker to make than natural casings, and they have the added advantage of being uniform and consistent because they are manufactured. They can be shaped and packaged in a way that best suits sausage manufacturers – it's hard to get intestines to do that! There are also fibrous casings that are made from plant cellulose; dissolved wood pulp is treated with a couple of chemicals in order to isolate fibres (the same that are used in rayon). These casings are built for strength and are often used to pre-cook frankfurters, but are then peeled off before the sausages go to the consumer. There are also plastic casings, which are often sold as high-barrier casings because neither water nor smoke can generally permeate them. Some of these plastic casings have been developed with a layer of smoked flavour on the inside so that the sausage develops a smokey flavour while wrapped in its plastic tube, without any actual smoke needed. In short, the desire to move away from intestines has resulted in the development of more resource-intensive processing methods for manufacturing casings.

The tweaking of the sausage-making process doesn't end with the casing. The surface mould is no longer left to chance. Moulds are now inoculated directly onto the surface of the sausages, and then they are placed in a room with the ideal humidity and temperature to let that mould grow at a calculated and controlled rate of growth. Once the layer is formed, the humidity is reduced so that drying of the sausages continues. The less moisture in the sausage, the longer its shelf life. Traditional mould-ripened sausages would have to dry for a minimum of 10 to 12 weeks, whereas sausages that have had starter cultures added take half this time.

It's not just some salted organs stuffed into an intestine anymore. Rules dictate the exact percentages of meat, fat, water, fillers, spices and additives. Science has made sausages safe to eat and fast to make – two things that can often be contradictory. Consumer pressure has pushed to have the 'yucky' stuff taken out of them (phew, no more rectums) and instead there is a long list of additives (yum, E310). Sometimes I think we may have missed the point of this whole sausage thing.

As Western societies got wealthier and food became more abundant, we got fussy. We transformed a perfectly good product designed for using up waste into a premium product containing quality cuts and trimmings (all the bits left over after the steaks, roasts and other cuts have been removed) tossed with herbs and spices, Bramley apples, cranberries, truffle oil and caramelised onions. OK, so maybe I'm describing one end of the sausage spectrum, and lots of them do still use the less desirable cuts and trimmings that can't find a purpose elsewhere, but for the most part, we are completely underutilising a significant portion of the animals we slaughter.

It's an offal waste

In the UK alone 940 million animals are killed each year for human consumption; 850 million of them are chickens. Each of those animals is inspected thoroughly to look for signs of disease, contaminants and signs of animal abuse. Many of the animals will be rejected because they are deemed unsafe for human consumption, and many of those will be rejected (wholly or partially) because of aesthetics. Much like astronaut carrots, there are natural things that occur in meat, from discolouration to abscesses, which pose no threat to humans but make the meat unappealing. Between 2012 and 2014, for example, 190,000 animals were rejected because of the presence of a tapeworm cyst, which poses no threat to humans whatsoever. I agree that a cyst in your roast isn't terribly appealing, but does it require the rejection of an entire animal?

Then there are all of the parts that as a society we once wholeheartedly ate in dishes such as humble pie (made from deer offal) and elder (cooked cow's udder), which we now reject. Though I have to say that coming from Canada to the UK, Britons (and indeed Europeans generally) do embrace offal more than North Americans; haggis (sheep stomach stuffed with liver, heart, lungs, oats and spices), faggots (minced pig offal), steak and kidney pie and luncheon tongue are all still alive and well in the UK.

When we raised our own animals and slaughtered them ourselves in Canada, we certainly didn't waste very much. My mother served jellied cow's tongue at my eighth birthday party and had my guests been any other than fellow farm-kids my social life would have been over right there and then. None of them ate it, but at least they didn't ostracise me. It's amazing how taking an animal's life along with the responsibility of disposing of its carcass can change one's attitude towards meat. However, I wouldn't say that animal organs are part of my everyday diet these days – I've gone soft. Yet these are safe and nutritious parts of the animal.

To make waste matters worse, a shocking amount of premium muscle meat that does make it into consumers' houses gets thrown out because people get confused about 'best before' and 'use by' dates, or they forget about it in the back of the fridge. Globally, we throw 570,000 tonnes of fresh meat away each year – the combined equivalent of 12 billion animals.* It is shameful on every level.

If, as a society, we are going to be fussy about largely only eating muscles, then we should use every last bit of muscle available on that animal, right? In the 1950s Japanese fish processors realised how much meat was left on the bone during normal fish processing, which launched a new era of machinery that could mechanically separate meat from bones. The first machines were developed for poultry and fish as the bones are small and soft so separation could be achieved by shear force between a rubber belt and a steel drum. However, by the 1960s and 1970s it was estimated

---

* This figure is from the book *Farmageddon* (Bloomsbury), which is a must-read for any meat eater.

that over two million tonnes of red meat was also being left on the bone, so the technology evolved to cope with large animal carcasses. Now the process takes the bones and even partial or whole carcasses (in the case of chickens) and applies shear force under pressure to get every last possible scrap of meat off the bone. The product is known as mechanically separated meat (MSM). If it is done under high pressure, the bones can essentially be pushed through a sieve, which breaks down the muscle fibres, creating a gelatinous pink paste that looks a lot like the inside of a hot dog. And there's a reason for this. This is exactly what is on the inside of a hot dog. The first ingredient in Oscar Mayer turkey frankfurters is mechanically separated turkey; same with Ye Olde Oak American Style Hot Dogs that you can find here in the UK, except it's chicken.

The process can also be carried out under low pressure, which leaves the muscle fibres intact, giving it the look of a fine mince and the characteristics of fresh meat. After the bovine spongiform encephalopathy (mad cow disease) in the 1980s and 1990s, MSM for anything other than pork and poultry was stopped. There were concerns that some of the body parts, particularly spines, put through the process could carry the misfolded proteins (prions) responsible for transmitting the disease. The UK continued to process beef bones, but only under low pressure – a process better known as desinewed meat. However, in 2012 the European Commission ruled that this was still MSM and it should be stopped immediately. It caused an estimated loss of £200 million for the UK meat industry. The industry argued that the low pressure doesn't pulverise the bone in the same way that MSM does and therefore doesn't present the same risk, but they lost the ruling. Beef MSM isn't allowed anywhere to my knowledge and in both the US and the EU, any MSM must be clearly stated on the label and most products have restrictions as to how much MSM can be added.

Many people aren't keen on the idea of MSM. There are concerns that this isn't the healthiest of products. In 2013,

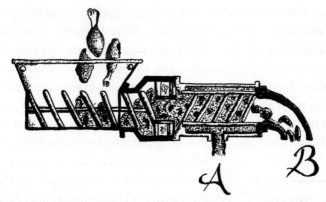

Figure 5.1 Auger-style mechanical meat separator. Carcasses go in through the top and then through mechanical shearing where meat and connective tissue is pulled off the bone. A paste of recovered meat and tissue is pressed out at A, while the waste bone is pushed out of B.

EFSA responded to growing public concerns and carried out an analysis of the public health risks of MSM from pigs and poultry. They found that MSM is usually rife with microorganisms, more than is found in a similar quantity of fresh meat or mince. There are a number of reasons for this. People tend not to treat carcasses with the same respect as they do a sirloin steak. The handling of carcasses bound for MSM need to be treated with the same best practices as any other meat in order to avoid contamination. Then there is the high-pressure process itself, which breaks down cells, releasing a plethora of nutrients for the microbes to thrive on. Finally, due to the fine texture of the product, there's considerable surface area for microbial growth. There are regulations that ensure batches are tested for pathogenic bacteria and MSM must be chilled and used right away or frozen immediately to slow the growth of any microbes. It can also only be used in cooked products. It turns out though, that if it's handled like any other meat product, MSM doesn't run a higher risk of contamination than fresh meat or mince and therefore poses the same health risks.

Including MSM in products like sausages and chicken nuggets, does, however, change the ratio of nutrients compared with using fresh meat alone. Because some bone and a lot of cartilaginous material is included in the recovered meat, MSM tends to have much higher levels of calcium, cholesterol and fat and lower levels of protein. This in itself isn't a health risk, but it changes what a consumer is perhaps expecting from a 'chicken' nugget. Considerable research has gone into looking at how much MSM can be added to products before people notice and before specified limits of things like calcium are exceeded. You can add as much as 50 per cent MSM to chicken breast to form nuggets before anyone complains and at 40 per cent nobody even notices. While packaged products in the grocery store require MSM to be labelled, restaurants and catering services don't need to specify whether it's in their products.

In 2012, there was uproar when Americans were made aware through an ABC News series that lean finely textured beef (LFTB) was added to approximately 70 per cent of minced beef sold in US supermarkets. Unlike MSM, LFTB doesn't involve any bones. The process takes the trimmings and warms them up in a gigantic mixing bowl that spins. As the trimmings get warm, the fat starts to liquefy and separate from the meat. Many of you will have eaten a piece of meat where there is a portion that has some fat and meat mixed together and the meat content is so low that it really isn't worth digging about to try and recover it. You cut the whole piece off, fat and meat, and push it to the side of the plate. Now imagine that for an entire carcass. All of those little bits of meat add up. This method helps to recover them without any chemicals or enzymes, just heat and spinning.

The product that is left is very lean beef bits. To this point, I think most consumers would be OK with it. However, they then puff the beef bits with ammonium hydroxide gas, which raises the pH of the product and kills any bacteria. This is when consumers freaked out. Ammonia is that stuff that you keep under the kitchen sink or in a high-up cupboard out of the reach of children because it has poison labels on it, isn't it? Well yes, but it comes in many forms and we produce it within our own body when we break proteins down. The liver then converts it into urea and we pee it out. So although ammonium hydroxide doesn't sound very nice, it is used in a lot of food processes and is globally accepted as being safe. However, when I discussed this process with analytical chemist Professor Richard Evershed at the University of Bristol, he queried whether this process would increase the nitrogen content of the product, potentially making it seem as though it had a higher protein content than it really did, which could be nutritionally misleading. Of course, we were researching food fraud at that point and we were suspicious of everything! However, I did have a look through the literature and I couldn't find anything associated with meat, but there was a study showing how nitrogen binds to wheat fibres when wheat is treated with the gas. An area for further research perhaps. Not all LFTB is treated with ammonium hydroxide; sometimes citric acid is used. The LFTB is handled and inspected just the same as any other meat product and it can then be added to minced beef products to increase the protein content. Because it is 100 per cent beef it doesn't need to be mentioned separately on the label, and I think it is this aspect more than anything that people got upset about.

And on this point, I think both consumers and industry could do better. As consumers, most of us eat processed food and we rely on the food industry to provide it. We are then shocked when we discover things in our food that we didn't expect to find. Fair enough. But we can be a bit of a tough crowd. If I can play devil's advocate for just a moment using sausages as an example: we don't want to eat offal, so the industry stuffs more muscle meat in. But muscle meat is worth a lot more as premium cuts, so they find ways to recover more meat from the carcass, reducing food waste in the process and increasing the value of the slaughtered animal.

Then we're told fat is bad for us and so we want leaner products. The industry develops a way to make products leaner by adding LFTB (in beef sausages, anyway). Consumers find out, freak out and brand it 'pink slime'. The industry loses millions.[*] I'm not a marketing expert, but surely if the meat industry had been open about the process and branded it themselves as a means of reducing waste, they would have won over more consumer acceptance? Consumers would be less worried because they would have known this was happening. Situation defused.

Consumers eat a lot of minced products – in fact, 44 per cent of all beef consumed in the US is mince. I don't think we should be surprised when the industry finds ways to include perfectly useful meat in these products, but I also feel strongly that we have a right to know it's there.

Fresh, plumped, formed and cultured

When I did research on food fraud, it became clear to me that sausages and other minced products could hide all manner of adulteration. Undeclared meats were being minced up together and nobody was any wiser until the horse-meat scandal, when authorities started to look more closely at authenticity. Like all foods, the more highly processed the meat, the greater the opportunity for fraudulent substitutions. However, we tend to point fingers at the product rather than the criminals responsible for this deceit; don't blame the sausage!

I was less aware how whole cuts of meat, items that aren't considered to be processed, are manipulated. Here is the crash course.

---

[*] Except for the US$1 billion that Beef Products Inc was awarded in June 2017 as part of a defamation lawsuit against ABC Network for their series on LFTB and concerns that it wasn't identified on labels.

Tough meat can be tenderised. We do this at home with a mallet-like thing; on a commercial scale a series of blades or needles perforate the meat repeatedly, breaking up the muscle fibres and connective tissue. This is mechanically tenderised meat and there's no hiding that this has happened. Safety issues have been brought up recently, however, because as the needles or blades drive down through the meat, they can transfer bacteria on the outside of the meat to the inside. The inside doesn't usually get as hot during cooking and therefore might not reach adequate temperatures to kill off any pathogens. As of June 2016, all beef mechanically tenderised with blades or needles (pounding is excluded) must be clearly labelled in the US. In the EU, it's enzymatically tenderised meat that must be labelled. Enzymes derived from plants such as pineapple and papaya are very good at degrading muscle protein rapidly. The enzymes are injected into the meat through cuts.

Fresh cuts of meat can be enhanced, injected or just generally plumped up (sounds a little like Hollywood). The process involves injecting, soaking, massaging or tumbling meat cuts in various solutions that improve their taste, tenderness, juiciness or colour. Sort of like a meat spa! I'm not referring to meat that is swimming in some marinade sauce – that's rather obvious. I'm talking about cuts of meat that look as if they have just been cut and packaged. The ingredients in such solutions include water, salt (added for flavour and juiciness), phosphates (for juiciness and antioxidant properties), organic acids (which improve shelf life and enhance colour) and antioxidants (which improve shelf life and retain colour).

This procedure hit the media in 2001 here in the UK when an investigation by the FSA found that chicken breasts being sold to the catering trade contained undeclared hydrolysed protein.* This hydrolysed protein helps the

---

* This is just protein that's been broken down into smaller units called peptides.

breast to retain water, which inevitably leads to a juicier breast, but also means that the consumer is buying a lot of undeclared water. Salted meat carries a lower import tariff, so importers were bringing in brined chicken from Brazil and Thailand and in order to make it palatable, water needed to be added. The first problem was that most of the products didn't declare this hydrolysed protein and added water on the label. The second problem was that some manufacturers were using proteins that weren't from chicken. The actual addition of these things isn't illegal; it's perfectly acceptable to say 'chicken breast with added pork protein'. The point is, it must be declared. In 2010, the UK FSA set out guidelines requiring that any 'added ingredients in meat products which have the appearance of a cut, joint, slice, portion or carcase of meat' be labelled prominently on the packaging. In June 2016 the USDA Food Safety and Inspection Service (FSIS) published a similar rule requiring that the details of added solutions be printed on the packaging. However, I'm not entirely sure how many people would look at an ingredients list when picking out a pork chop from the supermarket, as one would expect that it contains, well ... pork. Give it a try, though – you might be surprised at what you find out.

The food industry has also become more sophisticated in its ability to form meats from smaller cuts, known as formed meat. Many people are familiar with formed ham – the sandwich-sized homogeneous cuts of pink meat that have clearly not been carved straight from a leg bone. The formed rounds of lamb that rotate slowly in kebab and shawarma houses everywhere aren't pretending to be anything but a highly compressed and well-seasoned mix of meat bits. However, technology has become so good that it is now possible to take a few smaller cuts of meat and make it appear as though it is a single cut of meat, such as a filet mignon. These smaller cuts aren't necessarily lower grade in any way; they are just smaller than a single portion size. The smaller cuts are glued together using the naturally occurring enzyme transglutaminase, which encourages proteins to bond, just as in the blood-clotting process. Commercially, transglutaminase (aka meat glue) is produced by fermenting bacteria, or it is isolated from the waste blood from slaughterhouses. The idea is to use a waste product (blood) to turn lesser-value cuts into higher-value cuts. The problem is that this is once again a grey area. In most countries it must be clearly stated on the label if a meat has been formed, and this came about largely because of safety issues. By sticking two pieces of meat together, the outside surfaces get put into the middle of the cut. Just as in mechanical tenderising, this means that any surface microbes have now been transferred to the middle of the piece of meat, posing a potential health risk.

Finally, the future of meat processing has now gone to the extreme. Our increased consciousness of animal welfare and desire to eat only very select pieces of the animal has pushed us in the direction of cultured meat. Muscles can now be grown from cultured stem cells and reared and exercised independent of any living thing. In 2011, as cultured meat was still under development and everyone was looking to it as the potential answer to the omnivore's dilemma, scientists at the University of Oxford and the University of Amsterdam explored its environmental impact. They conducted a life-cycle assessment comparison between cultured meat and meat conventionally produced in Europe. They found that cultured poultry had 7 to 45 per cent less energy use, all cultured meats had 78 to 96 per cent lower greenhouse gas emissions, 82 to 96 per cent less water use and, not surprisingly, 99 per cent less land use. On paper, it seems like a no-brainer. In 2013, the world's first test-tube burger was introduced with an estimated R&D price tag of about £290,000 (US$375,000) and it tasted like ... well, let's just say it was a bit of a letdown.

Those pursuing artificial meat were not thwarted though. In 2017, Memphis Meats announced that they had developed cultured chicken and duck and that it had passed the expectations of some taste volunteers who declared that it tasted like chicken (doesn't it always).

Cultured meats are likely to enter the market within the next five years, bringing with them a whole new set of questions. On the plus side, animals are not being killed for meat production and some of the less marginal land that was used for rearing animals could be put towards growing other crops. It will undoubtedly present new dilemmas for people who refrain from eating meat for environmental and/or welfare reasons. It will trigger discussion among different religious communities and potentially invite new opportunities to commit food fraud. The reality of regulating cultured meat is already dawning on governments, who are no doubt tossing the responsibility from department to department like a hot potato. One has to think that the development of meat without animals has to be one of the biggest changes in meat processing since humans started cooking. Whether it is a step forward remains to be seen.

CHAPTER SIX

No Added Sugar

Sugar is a carbohydrate, as are starches and fibres that can be found in all fruits, vegetables, grains and dairy – in fact in pretty much everything. Carbohydrates are an important energy source, particularly for the brain. Like all macronutrients, carbs are broken down by enzymes in the alimentary canal and absorbed through the small intestine where they enter the bloodstream. We classify carbohydrates as being simple or complex based on how quickly the body is able to do this.

Carbohydrates are such critical sources of energy for humans that we are particularly efficient at using and storing them. And it is for this reason that carbohydrates, and refined sugar in particular, have become the enemy of the modern diet. Globally, obesity has more than doubled since 1980. In 2014, 39 per cent of adults over 18 years old were overweight and 13 per cent were obese. Nine of the ten most obese countries in the world are in the Pacific Islands. The Cook Islands top the list with over 50 per cent of the population considered obese. United States is number 12 at 35 per cent; Canada, Australia and the UK are 25th, 26th and 27th respectively at 30.1, 29.9 and 29.8 per cent obesity rates. With this increase in obesity, there have been increases in chronic diseases such as cardiovascular disease, cancer, Type II diabetes and other metabolic diseases. In 2008, the WHO reported that chronic diseases had surpassed infectious diseases, such as malaria, tuberculosis and HIV, as the number one killer globally. This is partly because fewer people are dying of infectious diseases, but mostly because more people are dying of chronic disease. There's no mosquito or dirty needle to blame, just a little genetic predisposition and bad food and lifestyle choices.

But this is not new to you. We hear about diet so often in the news that it can be incapacitating. Eat more of this! Eat less of that! Avoid all fat! No, just avoid saturated fat! Oops, got it wrong again, it's just the trans fat you need to avoid! Antioxidants, superfoods, complex carbohydrates and low GI foods are the way forward! We are nutrient obsessed, and yet when I prepare dinner, I'm thinking about what I fancy eating, what flavours will go with what, and whether I've got combinations that will look pretty on the plate. I'm not thinking about whether it will help me to achieve my daily recommended intake of a, b and c, or exceed my recommended intake of x, y and z. What would be the pleasure in that?

We are in a diet-based crisis and we are all looking for a smoking gun. Sugar, fat and salt have all had their turn at the end of the barrel. And yet, these are considered to be the tools of the trade for the food giants who are luring us into their latest snack products. In his enlightening book *Salt Sugar Fat*, Michael Moss explores how food manufacturers have been exploiting these three ingredients in different ways to get us addicted to highly processed foods, and it's apparently working. In 2014, consumers spent £292 billion (US$374 billion) on snack foods. Europeans and North Americans were the top snackers, spending £130 billion ($167 billion) and £97 billion ($124 billion) respectively. Globally the market is growing about 2 per cent each year, but in regions such as Latin America, this growth is more like 9 per cent.

The industry has seen the snack opportunity and developed a diversity of snacks in response. Personally, my kryptonite is potato crisps. I will have a well-balanced gigantic salad with no fewer than a dozen vegetables, probably some nuts and seeds to top it off. But later that evening those fried (or baked), salt-coated carbohydrates call to me from the hard-to-reach cupboard. Sales of salty snacks are worth about £21.6 billion ($27.7 billion) in North America, so I'm not alone. In Europe, however, the number one seller is sugar, with sales worth £36.2 billion ($46.5 billion), served in its different forms of sweets, chocolates and chewing gum.

If you've ever gone for any length of time without a lot of sugar in your diet, you will know how habituated we have become to its taste. After a no-sugar stint, grapes can taste uncomfortably sweet. The same happens with salt. We are finely tuned to identify and enjoy these flavours because over a couple of million years of evolution, they have served us very well, telling us when food is good and when it's not. We can taste sugar because it helps us to identify carbohydrate-rich foods – sweet receptors are the tool of any plant-eater seeking out starches and other carbohydrates. An obligate carnivore like a cat has lost the gene that codes the sweet receptor because it doesn't provide it with any information about the quality of its protein-based meal. Likewise, our dedicated sensors for salt help us to seek out essential minerals. Sensitivity to a sour taste lets us know when things have gone 'off'. It's recently been discovered that we can also detect fatty acids, which help us discern those satisfyingly rich sources of energy that also help us absorb fat-soluble vitamins.

While our early hominin ancestors probably had all of this sensory machinery in place, their encounters with sweet, salty and fatty would have been relatively rare. Fruit would have been seasonal and meat would have been hard earned. Now we're eating them both all the time and we've established a new baseline level, particularly for our children, which can only spell trouble.

Let me explain what I mean by that. When I was a child, I hadn't even seen a soft drink, let alone tasted one, until I was probably eight or nine years old. Then, for whatever reason, we would bring these fantastical drinks into the house for the week around Christmas just in case visitors stopped by. Cream soda, with hints of vanilla and a shocking pink colour, was my favourite and I would have one glass and then be sent outside to play because otherwise I would be bouncing off the walls. When I watch what children eat these days, they have probably had the cream soda sugar equivalent by breakfast. When I travelled to work by train, I would see teenagers on their way to school drinking energy drinks and passing around sweets on a daily basis. That's their starting point. Then they have a day filled with convenient lunchbox-appropriate snacks and sugar-based rewards for their achievements. When they come crashing down we placate them with more sugar. We are worried about our kids getting addicted to drugs? Well guess what, they already are. And we're the dealers.

There's a lot to be said in terms of salt and fat and how they have been added, removed and re-added to processed foods over and over again. I recommend having a read of Michael Moss's book. But here I want to focus on the one that I worry most about: sugar.

The carbohydrate expansion

Our early ancestors would have had very little sugar in their diet and certainly no refined sugar. Fruit and honey would have been the main foodstuffs with higher sugar content and these would have been very seasonal and limited. The fruit would have contained far less sugar than what we buy from the supermarket these days. It has taken us many centuries to develop varieties of fruit with more of the sweet bits and less of the annoying bits. Bananas, for example, had big hard seeds in them and far less of the fleshy part we enjoy now. Peaches were first domesticated nearly 6,000 years ago in China and they were about 25mm (1in) in diameter, mostly made up of the stone, and had more savoury overtones than sweet. A seventeenth-century painting of a watermelon reveals that it has only small

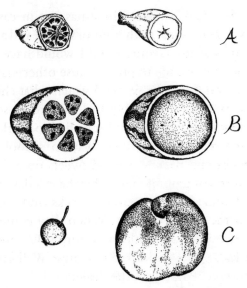

Figure 6.1 Fruits have changed over decades of breeding, seeking the sweetest flesh. A: bananas used to have large seeds (left) compared with the bananas of today (right). B: watermelon in the 17th century (left) had far less of the sweet pink flesh we covet than today's watermelons (right). C: ancient peaches (left) were small and savoury compared with the fruit of today (right).

pockets of the pink flesh we love, and it's been within my lifetime that the big black seeds have been reduced to virtually nothing.

Even honey, which has been harvested for at least 10,000 years, would have been a more daring endeavour with less reward; it's taken humans hundreds, if not thousands, of years of breeding to get such proficient honey producers as the modern honey bee.

Sugar cane originates from south-east Asia and it was probably originally eaten as it still is in many of these regions – hack it with a machete and chew, spitting out all the fibrous bits that are completely inedible. In its raw form, it's a lot of work.

Around 3,500 years ago sugar cane started to migrate out from south-east Asia to other parts of the world, carried by traders. Sometime around 2,000 years ago, people figured out that sugar crystals could be acquired from the cane juice. Like many great discoveries it was probably an accident, in that someone left a dish of sugar-cane juice to evaporate in the hot Indian sun, only to discover later that sweet crystals were left behind. A processing method was discovered. The crystals were much easier to consume than chewing on a big old piece of tough grass and could be stored and transported easily.

Sugar travelled the world. It reached Europe as early as the first century AD. By the seventh century the Chinese were journeying to India to learn more about their refining processes and technologies. In the ninth century Al-Andalus (the medieval Muslim territory that is now Spain and Portugal) became a centre for sugar production and the Arabs built mills and refineries right on the plantations. By the twelfth century Venetians were exporting sugar from Lebanon to Europe. In the fifteenth century the Spanish and Portuguese were introducing sugar cane to their newly conquered territories, including Brazil, the world's largest sugar producer today. In the late sixteenth century the Dutch took sugar cane from South America to the Caribbean islands. It was a fruitful crop, but producing it remained a laborious process, and so slaves were brought in to work the plantations. Sugar got cheaper and soon became more accessible to everyone. By the middle of the eighteenth century, sugar was more popular than grain in Europe.

Just over three centuries ago, sugar infiltrated the European menu. It offered a new opportunity for preserving fruit in the form of jams and jellies, it could be added to cream for desserts and it opened up a whole new world of baked goods. It was enjoyed in tea and coffee and yet it was still expensive enough to mean that sugar, like spices, could be used as a display of wealth. Truly glamorous dinner parties of the aristocracy would include a spread of sweet items. Sugar changed the way people ate and drank. In 1815, refined sucrose consumption in England was 6.8kg (15lb) per person per year. By 1970, this would rise to 54.5kg (120lb).

Today, sugar is still usually milled near the cane fields as it isn't terribly economical to transport all of that excess fibre around. The stalks are washed and cut into shreds with a machine and then the shreds are passed through huge rollers that press the juice out. While the leftover cane fibres usually go to fuel the mill, the cane juice is sent off to have any impurities removed. Calcium hydroxide $(Ca(OH)_2)$ suspended in water, known as milk of lime, is added to the cane juice and then carbon dioxide is bubbled up through the mixture. The calcium finds a better partnership with the carbon and forms calcium carbonate $(CaCO_3)$, the main constituent of eggshells and seashells. With its calcium partner gone, the hydroxide joins with free hydrogen atoms to form water molecules. The calcium carbonate attracts things like plant wax, fats and gums, which also got squished out into the juice, and pulls them down with it as it settles to the bottom of the container and is then filtered off.

The sucrose solution is then evaporated under vacuum so that it boils at a much lower temperature in order to prevent the sugar from caramelising. This is done in multiple stages and with each step the clear juice becomes a thicker rich brown syrup. The final evaporation stage to remove the last bit of water is also done under vacuum but pulverised sugar is added just beforehand to seed crystal formation. The mass of crystals is then sent to a centrifuge where it is spun to remove any remaining water. What is left is a dry, golden raw sugar that is between 96 and 98 per cent glucose. The remaining 2 to 4 per cent is a film of molasses coating the crystals. The molasses is made up of sugar, water and other plant materials and minerals that weren't extracted by the calcium carbonate. The raw sugar can be sold as it is, but usually it's sent on to a refinery.

The refinery mixes the raw sugar with a warm sugary water solution. This loosens the molasses coating from the crystal and then it is separated using a centrifuge. The golden liquid that is left over once the molasses is removed is then placed through carbon filters that absorb the impurities which impart the golden colour. The remaining clear liquid goes through the evaporation process again under vacuum and seed crystals are used to start crystallisation. If it's done properly, sugar crystals of the desired size are produced. The crystals are once again centrifuged to get rid of any remaining syrup and then they are washed and dried. Once dry, the crystals are passed through screens to separate them by size. They are then packaged up and shipped out.

Sugar cane is the source of 86 per cent of the world's sugar. The remaining sugar is produced from sugar beets – the sweet solution for temperate climates. The process for extracting sugar from beets is similar to that of cane sugar, with the exception that it doesn't require further refinement after the first crystals are formed. The product at the 'raw sugar' stage is pure white without the molasses coating.

The ability to refine sugar so efficiently and cheaply meant that it could become a widely used ingredient in food processing. It not only became a flavouring and sweetener, it became a tool. As we have already seen from previous chapters, it can be added to fermentation products, such as bread and sausages, to provide a rich carbohydrate resource for the microbes responsible for fermentation. Sugar is added as a preservative and flavour enhancer. It affects the texture and colour of products. In baked goods, it acts as a hardener, crystallising after baking. It is used in the preparation of sauces and ready meals to counter the acidity of other ingredients, such as tomatoes. It also counters the tartness of food additives, such as ascorbic acid and citric acid, which serve as antioxidants and microbial inhibitors. And while sugar was probably added initially to serve these functions, the reality is that if company A is adding more sugar to their pasta sauce than company B, company A will sell more pasta sauce. What does company B do? It adds more sugar.

The effect of sugar in the body is undeniable. If I am feeling particularly down, I will add a small amount of sugar to my tea and it makes all the difference in the world. Babies, who are blissfully unaware of sugar's bad reputation, will stop crying and endure discomforts better if they are given something sweet. Sugar makes us happy, having similar effects on the brain to cocaine and alcohol. We then start to crave that happy feeling, and sugar is cheaper than cocaine.

We don't even have to get that sugar hit from the white powder directly. White bread, pizza dough, potatoes and pasta are all sources of simple carbohydrates that our body converts into glucose very rapidly. Everything in our evolution as humans has said that this glucose hit is a good thing – that the fruit is ripe – and so we begin a vicious cycle.

When that spoonful of refined sugar in my tea hits my bloodstream, my body detects the rising glucose levels[*] and signals my pancreas to release insulin into the blood. This is a broadly acting hormone that alerts all the cells in the body that energy is on the way. The insulin disperses and binds to insulin receptors on cell membranes, which act like keys, opening the cells up to receive glucose. However, because my sugar is a simple carbohydrate, it has entered

---

[*] Sucrose (sugar) is converted by enzymes in the small intestine into both glucose and fructose, which are then absorbed into the bloodstream. Glucose will be used directly by cells in the body, while fructose is first sent over to the liver where it is converted to glycogen and triacylglycerides (fats). It can then enter the same biological pathways as glucose.

the bloodstream so rapidly that the pancreas panics a little
and sends an equal surge of insulin. The insulin is telling
the body to clear the glucose quickly from the bloodstream
so the excess glucose is converted into triacylglycerides (fat)
by the liver.

The spike of insulin has cleared the bloodstream of
glucose so efficiently, however, that I can start to feel a bit
tired, and I begin to crave another sugar hit. So, I feed it
with a biscuit. The glucose-insulin spike cycle repeats
itself. Had I not had that spoonful of sugar in my tea in
the morning and instead had a comforting bowl of
wholegrain porridge, then it would have taken longer for
my body to break those more complex carbohydrates
down into sugars and I would have felt fuller for longer.
The insulin response would not have been as great and
the cells would have been able to put the steady source of
energy to good use.

The challenge is that we are bombarding our bodies
with easily accessible sugars frequently. And just as our
sweet sensors and brain become habituated to certain levels
of sugar in our food, our body becomes habituated to
certain levels of glucose, or rather insulin, in our blood.
After being on the glucose see-saw for a while, the cells
don't respond to the insulin anymore, partly because fat
(the excess stores of glucose from meals gone by) is physically
interfering with the insulin receptor on the cell. This is
insulin resistance and it is the start of a journey towards
type II diabetes. Without insulin, the cells don't open their
channels to receive glucose and start to starve. They send a
desperate plea to the liver to release some of its glucose
from its stores. The pancreas produces more insulin in
response to the new influx of glucose, but the cells aren't
listening, and like any ignored organ, the pancreas
eventually gives up. This is type II diabetes.

The insulin spikes also stimulate the synthesis of leptin,
the hormone that tells the brain that the body is satiated

and to stop eating. With the increases of leptin in the body, the brain also becomes habituated and ignores the signal that it's full. So, we eat more; we get fat.

Sweet alternatives

The search for alternatives to sugar cane wasn't motivated by the desire to reduce sugar intake or reduce the global obesity problem. It began as a result of supply issues. In 1810, France's ports were blockaded by the British. Napoleon, having just received a successful entry from Appert for his food preservation prize, established the Sugar Beet Prize (worth 100,000 francs) to stimulate the production of sugar from beets. However, the French chemist, Joseph Proust, claimed the prize for extracting what would become known as dextrose, from unfermented grape juice. In 1811, the German chemist Gottlieb Kirchhoff discovered that he could convert starch into sugar by heating it with sulphuric acid (I've also read that hydrochloric acid was used). The acid wasn't used in the reaction, it simply acted as a catalyst. Kirchhoff's discovery was quite by accident, apparently. He overcooked a mixture of potato starch and sulphuric acid in his endeavour to improve porcelain manufacturing. He was left with a sweet syrup. Like Proust, Kirchhoff had found dextrose.

When the blockades came down in 1814 and sugar began to flow into France, these sucrose alternatives were all but abandoned. Kirchhoff's methods did find their way to America, though. In 1866, American chemists switched the starch source from potato to corn, and a hundred years later in the 1960s they replaced sulphuric acid with the enzyme glucoamylase. Corn syrup, primarily comprised of glucose, was created.

In the 1970s, corn syrup was taken one step further. Manufacturers found that some of the glucose could be converted into fructose using the enzyme glucose isomerase. The increase in fructose content improved the browning results and moisture-binding capability of corn syrup. Large-scale production of high fructose corn syrup (HFCS) began in 1978. It cost, at the time, half as much as sugar and so by 1984 100 per cent of the sweetener in Coca Cola and Pepsi was HFCS. It also found uses in baked goods, canned and processed foods, dairy products and confectionery. It became the primary ingredient in 'pancake syrup' (sacrilege to a Canadian I might add) and it became a clever adulterant of honey. HFCS is still widely used in food manufacturing.

The challenge with HFCS is that while it might present a cheaper alternative in food manufacturing, it is still high-energy and still generates insulin spikes. Health concerns over HFCS have been much debated, and claims have been made that fructose might be less satiating than glucose and get converted to fat more readily. Unsurprisingly, HFCS has been implicated in all of the same diseases as sucrose: insulin resistance, obesity and diabetes. To date, no study has been able to make a direct link between higher fructose consumption and a greater prevalence of these diseases though, which is why the WHO and most governments suggest that the recommended sugar limits include all sources of sugar, including HFCS.

A non-nutritive substitute for sugar, which maintains all of the functional characteristics of sugar, delivers the sweet sensation we've all come to expect and doesn't cost the earth, is the holy grail of food manufacturing. A whole host of natural and artificial sweeteners have been developed, starting back at the end of the nineteenth century with saccharin. One that ticks all the boxes still eludes the industry, however. Many are sweet and cheap, but they drastically change the texture, volume, colour, taste and shelf life of the foods to which they are added. Many synthetically produced sweeteners also degrade at temperatures that are frequently reached during normal cooking and baking, ruling them out for use in many processed foods. They also tend to leave a chemical aftertaste in the mouth that is incredibly unappealing, and if you will recall from Chapter 1, many of them were implicated in and then cleared of causing cancer.

With consumers wanting to see more natural ingredients on the label, considerable research has been put into finding a natural sweetener – these experiments have included stevia, thaumatin, inulin, tagatose and glycyrrhizin. These are compounds that are produced through metabolic pathways in plants. But they are generally found in small amounts in the plant and once identified, they have to be extracted and purified, just like sugar. Tagatose, for example, is found in small amounts in fruits and dairy products and it is 'generally recognised as safe' (known as GRAS within the industry). It is now produced commercially by breaking the milk sugar, lactose, into its constituent sugars, glucose and galactose. The chemical structure of the galactose is then altered using calcium hydroxide and high pH conditions. The result is a mixture from which tagatose can be isolated. It has 92 per cent of the sweetness of sucrose but with only 38 per cent of the calories. Because tagatose is not something humans encounter in high quantities naturally, it isn't absorbed very efficiently in the intestine. As a result, it doesn't generate the same insulin response as other sugars. Commercially prepared tagatose was approved for use by the US FDA in 2011.

Consumers tend to be dubious of new ingredients, however. As a result, lots of companies are exploring ways of modifying the sugar we're all familiar with in order to reduce its caloric value. Nestlé is thought to be developing methods to mill out the centre of sugar crystals as one approach. Nanotechnology, as you will read in Chapter 8, is also being applied to the problem. I have to admit, however, that part of me thinks that this would be the equivalent of milling out the centre of crystal meth. Aren't we missing the point?

Here's a crazy idea. Let's just change the sweetness threshold. Let's make those grapes and strawberries and other delicious fruits seem unbearably sweet again by making everything else less sweet. Can't we just admit that while sugar was fun for a few centuries, perhaps it didn't work out as well as we had hoped? Maybe we could turn some of the 22 million hectares of land used to grow sugar cane over to something that provides a greater diversity of nutrients instead. Perhaps we could find a better use for some of the £5.1 billion that is currently spent by the UK's healthcare system on diet-related disease. Smoking costs the UK's NHS £3.3 billion annually, and alcohol costs the NHS £3.3 billion, so food choice at £5.1 billion is the greater burden on the NHS budget. High taxes on sugar and warning labels akin to those on cigarettes have all been suggested. It seems a bit radical, I know, but then some would say milling out sugar crystals is radical too.

Preferences for sweetness seem to have geographical or perhaps cultural patterns. I ordered a Strongbow cider when I was back in Canada recently and I could barely drink it because it was so sweet compared to the British version. And yet when my family and I moved to England we noticed how much sweeter the curries were here. It doesn't seem to have been looked at, but I would imagine that there are also some temporal patterns. I would imagine the 'sweet spot', the optimal sugar content of food for children these days, is different from when I was a child, which was different again from when my grandmother was a child. The ubiquitous use of sugar has given us false perceptions of what food should taste like.

When I spent time in Ethiopia, the coffee was so sweet it made my brain hurt, but everything else they were eating was complex carbohydrates. Sugar, along with coffee, was being used as it should be, as a stimulant. But it's not food.

We have become so efficient at processing food that we have extracted the individual molecules that give us the most pleasure and added them into everything else that we eat. So much of what we do to food is to bring us closer to that sacred sugar. Juice is extracted from fruit so that the pesky fibre doesn't slow down the sugar rush. Fibre and nutrients are removed from milled wheat in order to be able to convert that starch into sugar faster in the body. Fruits and vegetables are bred to be sweeter. We are sugar junkies.

One of the things that may have helped us to last this long in this sweet world is our diverse diets. When our hominin ancestors were binge-eating seasonal fruit they probably weren't also hunting and fishing. It would have been rare that they would have been eating these rich carbohydrates at the same time as consuming fats – certainly not in the same meal, and maybe not even on the same day. Most of the meals we eat these days are a complex mixture of protein, fat and carbohydrates. There is considerable evidence to suggest that when we consume fat and protein with high glycaemic* foods, protein in particular slows the absorption of glucose into the bloodstream. Scientists are currently exploring whether this can be used as a quick-fix option (my words, not theirs) to address chronic metabolic disorders. A manufactured protein shot taken before each meal could help even out the glucose see-saw so many people are on.

It will take a strong political will to extract sugar from its prominent place in our food system, but as consumers

---

* The glycaemic index (GI) is a measure of how quickly carbohydrate-containing foods increase glucose levels in the blood. The baseline is straight glucose, which has a GI of 100. High GI foods send blood sugar rocketing while low GI foods keep blood sugar level.

we can, no, we *must*, support this by creating the public acceptance to enable the decision-makers. I'm not suggesting putting money towards finding yet another alternative, I'm saying cut it out. In the meantime, I think the single best thing I can do as a parent is – stop dealing.

CHAPTER SEVEN

The Convenience Conundrum

Initially I thought that the first convenience food was the TV dinner, introduced by Swanson in the 1950s in the US. However, I have recently been enlightened to the fact that the earlier predecessors of convenience food started decades before this. Custard powder made custard preparation far less fussy by the 1840s, and by 1889, Aunt Jemima pancake mix made consistently fluffy pancakes easy. However, convenience was not yet viewed as a priority because at the start of the twentieth century, most middle-class households in both Britain and the US had at least one servant responsible for all the cooking and cleaning. And there was a lot of cooking and cleaning. This was an era when people displayed their wealth through their food; they hosted frequent dinner parties, often with as many as 12 courses. Food was about pleasure. Convenience wasn't an issue because it was not the middle class who were inconvenienced. It was some poor woman sweating over the 'kitchener' (coal-fired kitchen range), who was probably earning less than the family spent each year on meat alone.

However, within the next decade, things changed for domestic servants. A number of labour laws improved the wages in factories, which attracted many young women out of the kitchen and onto the assembly line. Families had to work hard to retain the help they had, which meant providing better wages and better working conditions. And then legislation was passed in 1912 stating that everyone, from shop staff to household servants, was eligible for a half-day off each week, no matter what. This meant that the lady of the household, because women were

inevitably in charge of all things domestic, was suddenly
responsible for preparing at least one family evening meal
weekly.

Paper-bag cookery came to the rescue. The food was
placed into special bags, fastened and placed in the gas
cooker – that's right, a grease-soaked paper bag in a gas
cooker. Despite the potential risk of a kitchen fire, there
were many advantages. It took less time to cook, saving
money on fuel. It didn't smell, which I personally don't view
as an advantage, but apparently it meant the family could sit
in comfort without aromas sending children running into
the kitchen every two minutes to ask when dinner is.
Paper-bag food was also apparently superior in flavour and
nutrition. It was no-fuss in that no basting was required; so,
as long as the woman of the house was confident that the
whole thing wasn't going to go up in flames, she could
essentially leave it and do other things. Most importantly,
though, there were no greasy roasting dishes to wash up!
And, if the family was lucky enough to have a maid, she
could prepare the meal the day before and leave it in the
bag, so all the lady of the house had to do was pop it in the
oven and hope she didn't burn the house down. Convenience.

During the First World War, about 400,000 women
employed as maids in middle-class homes left domestic
services in order to help the war effort by working in
munitions factories and by keeping the farms going. Many
didn't return after the war ended as the hours, working
conditions and pay were far better (which is saying
something). This changed the life of many housewives.
They had been raised in homes that had always had a servant
and so there had been no need to learn how to cook. There
was a gap in society's culinary skills. For those who could
afford it, there was tinned food; Heinz tomato soup, for
example, was an imported luxury. It wasn't until the 1920s,
with more canning factories and better processing methods,
that most people could afford to enjoy the convenience of
tinned food. A quick dinner might have included some

tinned fish with tinned potatoes and vegetables – just heat and serve. Dessert would have been some tinned fruit and some Bird's custard. It took minimal time to prepare and zero culinary skills. It saved the middle-class housewife.

Other conveniences were introduced in the 1920s. Clear Pyrex dishes reduced the cooking time of casseroles and looked nice enough to go directly from oven to table, which meant less washing up. A Kellogg's Cornflakes manufacturing plant opened in Britain, providing a new quick breakfast alternative. Quick Quaker Oats were introduced, which were cut much smaller so they could be cooked simply by adding boiling water. And it was the start of the UK's nationwide love of snack foods. Frank Smith and Jim Viney started selling crisp packets out of Frank's garage, distributing about 1,000 bags a week, mostly to local pubs in London. However, by 1930, Frank had bought Jim out of the business and there were seven factories across the UK making Smith's Salt 'n' Shake crisps.

By the 1930s, the larder or pantry of the average British home contained pre-made sauces, such as HP sauce and tomato ketchup. There were biscuits, such as bourbons and custard creams, and there was a greater variety of tinned soups. There were indeed so many manufactured goods on the market that manufacturers had to go to great lengths to make their products stand out from the competition. They began enriching their products with vitamins and making health claims. Pre-sliced bread was introduced in the UK, which made the ultimate convenience food of the time – the sandwich – even more convenient.

It is John Montagu, the fourth Earl of Sandwich (1718–1792), who is credited with developing the concept of the sandwich. I can't speak for the credibility of the information, but it is said that he wanted a quick and easy meal for his supper that he could eat while at the gambling tables, and meat wedged between two slices of bread was perfect. Montagu popularised the concept among the English gentry and it became the food of choice as the men drank late into the night; it was the equivalent of today's post-clubbing kebab, pizza or shawarma. By the 1930s, savoury sandwich fillings were all the rage and upscale sandwich bars were opening up all over London as the 'fast food' of the day. The sandwich remains a go-to convenience food, with 6.4 billion of them consumed each year in the UK, making it the most commonly eaten meal.

The 1930s also began the introduction of a number of appliances that made life in the kitchen more convenient. About one-third of homes in the UK had electric power and so electric toasters and kettles were introduced.

The 1940s of course were about winning a war – and the problem was sufficient food rather than quick food. Food was not for pleasure; it was, once again, a fuel. Convenience was less important, despite the fact that a huge number of women, even those with young children, had entered the workforce and still remained responsible for the household. Rationing was in effect and imported food was extremely limited, so people were eating seasonally and buying whatever their local butcher, baker and greengrocer had in stock. Extravagant items, such as cocoa, were in short supply, and yet desperation drives innovation. The Italian pastry maker, Pietro Ferrero, came up with the idea of adding hazelnut paste and sugar to eke out the small amounts of cocoa that were available after the war. The paste mix could be formed into a log, sliced and served on bread. Ferrero called it 'giandujot' and launched it in 1946. His son, Michele, then took Ferrero's innovation one step further by making it spreadable, and in 1964 he launched Nutella.

Things weren't much different over in North America. In the US, five million women joined the workforce between 1940 and 1945. In both Canada and the US, women's branches of the army, navy and air force were formed so that women could take over positions that would free up more men to go to the front lines. Government-run daycare centres were established to enable more women to go out to work. Yet after the war many women were kicked out of their jobs to make way for the returning servicemen. Women with a secondary-school education, however, who had secured more secretarial-type jobs that men didn't want, managed to retain their jobs; more women, albeit only within a very specific segment of the population, began to work outside the home. Convenience foods, such as Minute Rice, Kraft Parmesan grated cheese, Betty Crocker's cake mixes, Reddi-Whip, Nestlé's Quick instant chocolate milk powder and an array of frozen foods, including pre-cut frozen chips, were launched in the US in response.

In the UK, the post-war reconstruction effort was so significant that women were encouraged to remain in the workforce. However, as always, none of these working women were exempted from their domestic duties. War had changed the way the British were eating. They ate four meals a day and grew half of the food themselves. Everyone ate seasonally. It wouldn't be until the 1950s that easily prepared meals would take hold in the UK.

The 1950s was truly when convenient meals came into their own. The American company C.A. Swanson & Sons had started out making butter and margarine, but after the Second World War, they took advantage of the increasing popularity of frozen food and began selling frozen turkey, chicken and other meat. Their journey towards serving complete frozen meals, as I understand it, began with an abundance of unsold frozen turkeys – half a million pounds of them, to be exact. In 1953, Swanson had overestimated how many turkeys Americans would eat for Thanksgiving and they were left with a profusion of poultry and not enough refrigerated storage. They put the birds into refrigerated boxcars and as their frozen turkeys travelled the railways, the Swanson executives tried to think themselves out of a significant financial hole. The solution came from executive Gerry Thomas, who had been in Pittsburgh visiting the catering facilities for an airline. On the flight home, he came up with the idea that the very same trays that the airlines were using to keep food warm, could also be frozen and then heated. They could cook and slice up their abundance of frozen turkeys, add a side of vegetables, stuffing and gravy, and see if the American public was willing to eat turkey outside of Thanksgiving and Christmas, so long as they didn't have to prepare it. To market it, they paired it with the other marvel of the 1950s and it became known as the TV dinner.

Of course making the idea a reality wasn't necessarily straightforward. In 1953, Betty Cronin, a bacteriologist, was 21 and had recently been made Director of Product Development at Swanson. She was tasked with making the TV dinner safe, palatable and profitable. The challenge was that there are a number of different items on the tray – turkey, sweet potatoes, peas and stuffing – that all need to be tolerant to freezing and then be ready at the same time when heated. She tasted a lot of failures, but in the end they made something that was passable and Swanson sold about 5,000 of the meals that year. The following year, with Swanson executives finally supporting the idea with a nationwide advertising campaign, they sold ten million meals. These convenient meals retailed for just under a dollar, roughly the equivalent of a very modest meal at a restaurant at the time. The advantage was that these little trays of food could be eaten with your slippers on and in the company of the characters from *I Love Lucy* or *Father Knows Best*. They were sold as a revolution for the busy American housewife.

Well, Caucasian housewife anyway. If you have a look at the advertising from the era, everyone is white, despite the fact that more non-Caucasian women were in the workforce in 1950 (37.1 per cent compared with 28.1 per cent). The message in the ads is that if you're running late, don't worry because you'll still be able to feed your family on time with a Swanson meal! Did your husband bring home his really annoying golfing buddy without telling you? Don't worry, just pull another Swanson meal out of the freezer! Kids driving you batty during mealtimes? Just plunk them down six inches from the television with a Swanson meal and you can finally dine in peace. Oh, and because there's no washing up, you can pour yourself that gin and tonic even earlier! I honestly don't blame the 1950s housewife for buying into the convenience concept one bit.

The turkey TV dinner was also a revolution for the food industry. Not only was it a marvel in terms of developing a system where prepared food could be frozen and then reheated in the home, it was also proof that people were willing to eat turkey leftovers outside of traditional turkey-filled holidays. Most importantly though, it showed the industry that American consumers were willing to compromise taste for the sake of convenience – and this was pivotal. Hank Cardello, a former executive for some of America's largest food manufacturers, claims that this paved the way for a whole new way of thinking within the industry: 'If the consumer demands bigger portions, you don't raise the price; you use lower-grade ingredients and adjust the taste to a manageable level.' Food didn't need to be the best-tasting so long as it met criteria for convenience, price and portion size. Taste expectations could be managed.

Captivated by convenience

Tim Harford made a great point in a recent BBC news story. He wrote that despite the washing machine being held in high regard as an appliance of convenience, it didn't really save much time for housewives over the long term. Before the washing machine, people just washed their clothes less frequently. They used replaceable collars and cuffs, and donned pinafores and aprons, and just coped with dirty smelly clothes because there was no alternative. If housewives had stayed with that regime, the washing machine would have been great – one load per week in the machine and the lady of the house would have been done. However, because the machine made laundry easier, people no doubt became a little more reckless with their clothes, daring to stir the tomato soup without an apron, for example. Fashions also became simpler so that clothes could be mass-produced, making them cheaper. Instead of dirty clothes, people just had more clothes. The housewife went from scrubbing some good shirts for half a day each week to doing a load of laundry every other day. Those who couldn't afford the technology probably weren't all that worse off, except that everyone around them would have smelled flowery fresh with their frequently machine-washed clothes, tempting everyone to keep up with the Joneses in terms of modern conveniences.

Unlike laundry, though, a household couldn't just eat once a week, so the woman of the house had to put in the time every day no matter what. There also would have been no outwardly obvious advantage to serving up a home-cooked meal over a TV dinner – it's not as if the kids would smell better if they ate home-cooked food. However, over time, comparisons would start to be made: there would be neighbours who seemed to have more time to go for walks with the kids, or a friend who seemed to always have time to read the latest issue of *Good Housekeeping* or make a new dress for herself. When convenience meals came on the market, they gave women a little more time, which gave them options. And in the 1950s and 1960s, women were desperate to have more options: women were choosing to work outside of the home (and starting to argue for equal pay); they could go to university in mixed colleges (although it wouldn't be until the end of the 1960s and into the 1970s that they would be accepted into Ivy League schools in the US); and they could speak more openly about sex and take control of contraception for themselves. The option of not having to cook every meal from scratch was again about having choice.

Since those first TV dinners were served up, the average household has changed considerably. The number of women in the US workforce has increased substantially: in 1950, about 34 per cent of women were employed outside of the home, and by 2015 this had gone up to over 56 per cent. Numbers are similar here in the UK – 46 per cent of women were employed in 1952 and in 2012 this was up to 66 per cent. Beyond that, the number of single-parent families has more than tripled, particularly those of women who are raising children single-handedly without ever getting married. This was scandalous in the 1960s and therefore unheard of, but in 2012, 11 per cent of US families included single, never-married mothers. The 1950s 'homemaker' isn't at home any more. She's juggling a job or two, raising kids, keeping a home, balancing the books, baking for the school cake sale (or at least buying some cupcakes) and doing laundry in that fabulous time-saving machine of hers. And single dads are doing it all too! No wonder home-cooked food gets pushed down the priority list when there are convenience foods available.

However, a somewhat closer look at households reveals some other truths. People living with partners (married or otherwise) tend to share the domestic duties a lot more. Same-sex relationships apparently have been more successful in achieving this balance, but heterosexual partners are improving. Since 1965, fathers have, on average, tripled the amount of time they spend looking after their children (which is still only half the amount of time that mothers spend). They also spend two and a half times more hours each week doing housework and cooking (which, again, is still only half the amount of time that mothers spend). And if sharing the domestic workload doesn't make some free time in the weekly schedule, surely a trend towards spending less time at work does? In 1870, the average working week was 60 hours. In 1952, the average working week in the UK was down to 44 hours and people got an average of 16 days' paid annual leave each year. Now, the average full-time working week is even further down to 37 hours and there's a minimum annual leave of 28 days[*]. All summed up, both men and women have more leisure time now than

---

[*] Of course, this is what we are paid to work. It doesn't account for the unpaid overtime that everyone puts in simply to keep up in this culture of long hours that we have created for ourselves. It also doesn't consider longer commutes that many people make to and from work these days.

they did in the 1960s (at least on paper) – men have gained about six hours a week while women have gained about three. And yet, we are spending far less time in the kitchen preparing the food that feeds us. This would suggest that society's love affair with convenience food isn't born entirely out of necessity, but rather out of a conscious decision that preparing food is not a priority. This is fine, but I think it's very important to make this distinction.

Being busy seems to be a bit trendy, quite frankly, and the food and beverage industry knows it. Between 2011 and 2016, there was a 54 per cent increase in the number of global food and drink launches that made 'on the go' claims. Think about some of the advertising you see. A mud-faced mountain biker fuels up with a breakfast bar before careening down some forested trail. You can buy little perfectly portioned tubs of food that fit in your desk drawer so that you don't have to stop working in order to refuel. A quick-fix candy bar can solve your 'hangry' moment in the locker room at half time. The message is that we are a society on the go. We don't have time to stop what we're doing and enjoy a meal, let alone cook one! Though can I just add here that we do seem to have lots of time to stop and watch people preparing meals; at least one cookery show makes the top 30 viewed TV programmes each week.

Listen to people's conversations on the street: we (myself included) love talking about how busy we all are, second only to discussing the weather. Imagine that instead of saying how many hours they're putting in at work and how hectic home life is, someone instead spoke of their productivity at work, of making time to nurture and feed their physical and mental well-being, and of the satisfaction they had in making time for family, friends and community. It could significantly change the conversations we have with one another.

My personal experience has been that I enjoy being busy, but when I no longer feel as though I have control over how I spend my time, then I feel stressed. That is why we need periodic holidays from the various demands on our time, our schedules and routines. Eating is non-negotiable, so convenient food provides a means of regaining control over some time. Many people try to assert control over their time by multi-tasking, having more leisure time by choosing to zone out in front of the television (maybe even ironically a cookery show) while eating, for example.

So, whether people actually have more or less time now than they did 60 years ago is debatable, but the perception is certainly that we have less. We (particularly women) certainly seem to have more options for how we spend our time. And the rise of the ready-made meal (more commonly known simply as the ready-meal), with a global value of US$83.4 billion (£64.1 billion) in 2016,* would suggest that many of us are choosing not to spend it cooking. Even since the 1980s we have halved the amount of time we spend in the kitchen.

Ready-meals aren't only about time, though. Far more people live on their own these days. In 1950, about 9 per cent of households in the US were single people. By 2013, this had nearly tripled to 27 per cent. I know that when I'm on my own, my eating habits change considerably. I tend to graze more and eat lighter, more frequent meals. I would rarely cook a big meal for myself because I would be afraid I'd spend the next three days eating it for every meal in order to use it up.

People are also living longer and without as much assistance from family, so for elderly people living on their own, the

---

* This value comes from the report *Global Prepared Meals Sector: Analysis of opportunities offered by high growth economies*. But values vary depending on how 'prepared meals' is defined – for instance, whether it includes convenience foods like cereal bars. The World Bank estimates the global value of 'packaged food' to be $4.8 trillion (£3.7 trillion). Schmidt Rivera and her colleagues state that in 2011 the ready-made meals market was worth US$1.1 trillion (£850 million) (the paper is listed in the notes at the back, under Chapter 8). Another source estimates processed food and beverages are worth $4.6 trillion (£3.5 trillion). The point I'm trying to make with all of this is that it's worth a lot of money.

ready meal is a means of maintaining independence. Both of my grandmothers lived independently well into their late 90s and although they were both capable cooks in their day, preparing meals in the kitchen became more challenging for them. Ready meals provided them with more options.

The lifestyle of the modern household also means family members are often eating independently, grabbing food at odd times before they head off to another activity, rather than joining everyone in a family meal. Ready meals can provide some variety without having to prepare an entire roast dinner or pasta bake. And the variety they offer isn't just for single people. The frozen-food retailer, Iceland, is currently running an advertising campaign that shows families eating foods that they might not otherwise have tried. The exotic meals – from Spanish paella to sweet and sticky Korean pork ribs – come frozen, and all they have to do is pop them in the oven. They are selling simple culinary adventures that at a median cost of £2.20 (US$2.79) per portion, are cheaper than eating out and don't require a bunch of ingredients that one isn't likely to use again; the Chinese five-spice in the back of the cupboard is a classic example.

Ready meals are also perfectly proportioned with the nutritional information displayed right on the side of the box. This is perfect for the lazy calorie counter. Calorie-tracking apps make it easy to scan the product and add the nutritional data into the daily tracker. This is much easier than trying to figure out how many grams of each vegetable were in the salad you made for lunch. The food industry saw this opportunity in the 1980s. Stouffer's launched their Lean Cuisine line, ConAgra Foods launched Healthy Choice and Kraft launched South Beach Diet; you don't even need to count calories, so long as you are eating three square (in this case literally because they come in a square-ish box) pre-prepared meals a day, you will be within the daily recommended energy intake. Even some supermarkets now have their own brand of calorie-reduced ready meals; they offer instant portion control.

Ready meals might also be about skills. People turn to them not because they don't want to cook, but rather because they can't cook. For me, this is the most concerning reason of all. In the UK, the organisations Love Food Hate Waste and Mumsnet conducted a survey and found that over 60 per cent of parents with children over the age of three spend three hours or less a month cooking with their child. I was actually surprised it was as high as that. They found that a quarter of parents surveyed didn't feel confident about using up leftovers without a recipe to work from. If someone isn't a confident cook, then why risk the effort and time to make something that might taste absolutely horrible? You know what you're getting with a pre-prepared product. This also leads into the related argument that a lot of parents choose ready meals, such as fish fingers or chicken nuggets and chips, not only because they are fast and easy, but also because their children are more likely to eat them. I can't support this argument in the slightest as I firmly believe that children will eat what they've been raised to eat. I think that as a parent I have a responsibility to teach my child about food and expose him to as much culinary diversity as possible. If he knows how to eat properly, this has a better chance of ensuring his health and longevity than learning calculus does (not that I have anything against calculus, I just haven't used it once since I left school, that's all).

The cost of convenience

The UK seems to be a hub for ready-meal research. This is probably due to the fact that we are one of the world's largest consumers of ready meals, second only to the US. In 2015, the UK ready-meals market (chilled and frozen) was worth approximately £3.15 billion (the US market is approximately three times this). In 2006, 40 per cent of UK households were eating at least one ready meal each week; I can only predict that this number is higher now. The ready meal is no longer the occasional treat, but a regular part of people's nutrition. And yet ready meals have been linked to obesity, which indirectly links them to chronic diseases such as diabetes, cancer and cardiovascular disease. Even when lifestyle factors, such as exercise, stress and alcohol consumption are controlled for, people who eat ready meals regularly are at increased risk of obesity. There seem to be some nutritional costs of convenience, but what is fact and what is hype?

Before looking at the nitty gritty of nutrition, the first question is whether most ready meals are the right portion size. If we consider that we generally eat three meals a day, each of those meals should have the energy value of 500–700kcal, based on government-recommended guidelines and accounting for snacks and beverages. When I coast through Tesco's chilled and frozen ready meals, most seem to fit within that range: beef casserole is 567kcal and a roast chicken and pancetta bake is 652kcal. Quite a few fall short of 500kcal, making me wonder whether they would be enough for a full meal: Tesco's cottage pie is 468kcal and the Weight Watchers' frozen meals are generally under 400kcal. The University of Glasgow, in 2012, took a far more scientific approach than this and looked at the nutritional information of four of Britain's favourite ready meals (macaroni cheese, lasagne, cottage pie and chicken tikka masala), sold in the five largest supermarkets (Tesco, Asda, Sainsbury, Morrisons and the Co-op). Of the 67 meals they looked at, 32 didn't have enough calories to constitute a meal and 10 of them had over 700kcal. The researchers highlighted Tesco's chicken tikka and korma with pilau rice meal for two, as each serving has over 1,300kcal and 98 per cent of the guideline daily amount (GDA) for saturated fat. When I look it up now, five years later, the kcals have dropped slightly and the saturated fat GDA is down to 74 per cent, but it still has a long way to go. It shouldn't be assumed that a ready meal contains the right amount of energy for a meal.

Another UK study in 2012 compared 100 standard ready meals sold in the three top supermarket chains across the UK with 100 equivalent meals taken from top-selling recipe books of celebrity chefs. They averaged the nutritional content of each of these groups and compared them. Their results were not what I would have expected. Per portion, the meals prepared from recipes were significantly higher in energy than ready meals, with the median value for the homemade meals coming in at 604kcal, compared with 494kcal for ready meals. Of the ready meals, 18 per cent met WHO goals for the recommended ratio of calories from carbohydrates, while only 6 per cent of the recipes did. Ready meals won out on fat as well – 37 per cent met WHO goals, but only 24 per cent of recipes did. Most surprisingly, however, 56 per cent of the ready meals met WHO goals for fibre, while only 14 per cent of the recipes did. Where ready meals fell very short, however, was in salt content – only 4 per cent of ready meals met the guidelines, while 36 per cent of recipes did.

I think the point here is that we have to be conscious of what we're putting in our mouths, whether it's a favourite Jamie Oliver recipe or a ready meal; some are going to be energy-dense foods (often called 'comfort' food) while others are lighter meals. Some are going to be more nutritionally balanced than others as well. It's known as variety.

There have been concerns raised in the media regarding the quality of micronutrients in ready meals – vitamins, minerals and antioxidants. Some vitamins are most certainly lost during processing, through heat, or changes in pH or moisture, or even exposure to light. If we cook food at home or in a factory, these same losses will occur. The advantage of home cooking, however, is that we have more control. We can choose to steam vegetables to preserve more nutrients or cook them for less time. We are less concerned (and less regulated) about shelf life and food safety than a company that is providing millions of meals all over Europe. However, on the flip side of this, manufacturers have access to equipment that the home kitchen doesn't have. In recent decades food science has developed more methods of processing that decrease the loss of nutrients – ohmic heating, for example, where a passing electric current dissipates heat directly into food with less nutrient loss than traditional heating methods. Many manufacturers also fortify meals in order to restore any of the nutrients lost in processing, something unlikely to happen in home cooking.

The biggest nutritional differences between a home-cooked meal and a ready meal might be a consequence of ingredient choices. If the horse-meat scandal of 2013 taught us anything, it's that manufacturers are driven by bottom lines and reducing ingredient costs translates directly into increased profits. Unlike the home-cooked meal, there is no guarantee of the quality of the ingredients listed on that label. Cheaper oils, lower-grade vegetables and meat cuts could have nutritional consequences that we're not aware of. But more concerning is, of course, when what is listed on the label isn't in fact what is contained within the pack, and the more complicated the food network, the greater the opportunity (intended or not) for this to happen.

Additives are another concern when it comes to pre-prepared meals. What chemicals are the industry having to add to the meal in order to give it the flavour, appearance and shelf life consumers are looking for? When I looked at a ready meal of spaghetti bolognese, there was nothing on the ingredient list that I wouldn't put in my home-made version, except perhaps that they have used caramelised sugar to counter the acidity of the tomatoes, whereas I would probably just use plain sugar, and maybe not as much. The wheat flour has been enriched with vitamins and minerals – iron, niacin and thiamin – but this would be the case for my spaghetti as well unless I was grinding my own flour. Expecting to see more preservatives in a chilled meal, I examined two paella meals, one chilled, the other frozen. The only preservative is found in the chorizo sausage, and that's in both meals.

Most of the meals I looked at were very high in salt. A calorie-restricted chicken and lemon risotto meal that I came across had 23 ingredients listed for the cooked marinated chicken alone: three different types of sugar, stabilisers (carrageenan) and polyphosphate salts, which are both preservatives and emulsifiers; these would not be necessary in a home-cooked meal. There is also the oh-so-vague 'flavourings'. I came across a box containing two egg and cheese omelettes, which has to be one of the easiest things to whip up in under 10 minutes, and I can't believe this is even a product. The ingredients include 'vegetarian mature cheddar cheese', which I assume means it has been made with vegetable rennet, and the stabiliser xanthan gum, which is a thickening agent and also prevents the ingredients from separating. Once again, it's a mixed bag out there – while some meals seem to have an ingredients list not dissimilar to a home-cooked meal, others have included various functional additives and ingredients that you might not expect.

If there is variety among ready meals within a single store, it is no surprise that they vary across countries that have different food regulations and labelling laws. Comparing a spaghetti bolognese ready meal across the Atlantic leads to a few insights. The UK spaghetti is made of durum wheat and water, while the US spaghetti has gluten added on top of the flour, egg white (egg is not uncommon in pasta), and the emulsifying additive glycerol monostearate. However, the US spaghetti also has folic acid and riboflavin (added micronutrients), which UK spaghetti doesn't. The other main difference is that the US spaghetti meal has two types of salt (NaCl and KCl), as well as the preservative citric acid and the stabilising agent cream of tartar. The US portion size was smaller than the UK (213g versus 390g) and the two meals weren't too dissimilar nutritionally. The only glaring difference was that despite being the smaller portion, the US meal had twice as much sugar as the UK meal (14g compared to 6.7g).

The nutritional costs of eating ready meals are not as clear-cut as I expected them to be. There are a number of studies that suggest there are nutritional consequences of eating ready meals regularly, but, as with any food, there are some options that are better than others. Not all ready meals are created equal, but nor are all home-cooked meals.

But what about other costs? One of my biggest concerns about the ready meal is its environmental cost – between transport and packaging, surely the ready meal is an environmental nightmare. When you consider various environmental impacts in the EU, food consumption accounts for 20 to 30 per cent of the degradation. For eutrophication, which is the nutrient run-off from land into water bodies, food consumption (mainly production) accounts for 50 per cent of the impact.

In 2014, a group from the University of Manchester conducted a life-cycle assessment of Britain's favourite ready meal, a roast dinner consisting of chicken, vegetables and tomato sauce, and compared it with the same meal prepared at home. Both meals start out the same, requiring the rearing and cultivation of raw materials, packaging and transporting those materials for processing into ingredients. The ready meal then has the additional step of meal manufacturing, which includes transportation to the manufacturer, the energy and water used in manufacturing that food, waste associated with that manufacturing, and then packaging, transporting and storing that food for distribution. From there, the homemade meal and prepared meal have a similar journey in being transported to retailers, stored and then packaged up by the consumer and carried home. As one would expect, the overall impact of the home-cooked meal is less because it avoids all the additional transportation and cold-storage costs associated with the manufacturing process. However, there were some things that surprised me.

Ready meals prepared from fresh ingredients (rather than frozen) and then immediately frozen are more environmentally friendly than fresh chilled meals. The main reason for this is that chilled meals are usually kept in less efficient open refrigerated units at the supermarket rather than behind closed doors, such as in the frozen-food section. Once that ready meal is home, it's more energy efficient to microwave it rather than heat it in an electric oven: an electric oven has 6.5 times more global warming potential than a microwave. Home-cooked meals are better for the environment when cooked with gas rather than electric appliances. The most surprising result for me, however, was what happens when different sources of ingredients are used. If I were to make this meal at home, I would specifically try to source British ingredients. A manufacturer of the ready meal might be more interested in cost and therefore source the chicken from Brazil and the tomatoes (for the sauce) from Spain. In this scenario, the ready meal becomes the better option when it comes to five different measures of environmental impact, including global warming potential. This is because despite the greater transportation costs, chickens reared in Brazil have a lower environmental impact than British chickens, and British-grown tomatoes require electricity-heated greenhouses, while Spanish ones don't. Of course, the reason Brazil-reared chickens have a lower environmental impact is because they are kept in much higher densities and with fewer resources such as dry bedding. It might be an environmental win, but in terms of animal welfare it's not a good choice. And of course there are economic benefits to supporting British farmers, despite potentially higher environmental impacts.

Packaging is another area where you would expect to see some benefits of the home-cooked meal. Again, this depends on how the ingredients are sourced, and also on consumer behaviour. There is less packaging if the ingredients are bought from a butcher or greengrocer and if the consumer chooses not to put the loose vegetables in a plastic bag. But if they are bought from a supermarket, each ingredient will probably come with its own packaging. When these same ingredients are transported in bulk to the meal manufacturer, there is generally less packaging per volume of product. And in either scenario, the consumer will have packaging waste to deal with. As mentioned in Chapter 4, while it's a nice thought to reduce packaging for environmental reasons, there is then a greater risk of waste, which is the least environmentally friendly outcome of all. Food is a very complex issue to navigate.

Beyond all of these costs, however, I think my greatest concern with the increased popularity of ready meals is that we are at risk of losing a critical skill. If I might return to the clothing analogy that I brought in earlier (which is odd because I'm the least fashionable person I know), there was a time when the majority of women were skilled in making clothes. I personally don't know a single person who makes their own clothes now, or to my knowledge, has ever made a single item of clothing with the exception perhaps of a knitted scarf or hat. Clothes are mass-produced by strangers and we have been willing to compromise a tailored fit and quality for convenience and cost. The result is that few people of my generation (Generation X) would have the ability to whip up clothes to keep themselves decent and warm should the fashion industry cease to exist tomorrow. I would probably be forced to strategically wrap myself in a knitted scarf as this is pretty much the extent of my abilities. Clothing is an extension of our identity and reflects our culture, but it is also very functional in protecting our naked bodies. And yet we have consciously chosen to let that skill fall by the wayside.

Is it possible that we are heading in this same direction with food? Are the majority of people in developed countries going to hand over the responsibility of food preparation to others, and in doing so, lose the skill for themselves? That deeply concerns me.

It also deeply concerns some others out there. Jamie Oliver and his foundation have worked incredibly hard at trying to incorporate food skills into mainstream education. When he didn't have any success here in the UK, he began to work in North America. Jamie also tried to make home cooking fast and easy, first with his 30-minute meals and more recently with his 15-minute meals. I'm not sure whether even Jamie could get it much below 15 minutes though. The organisation Love Food Hate Waste is trying to help reduce food waste by improving people's confidence in the kitchen. People are more likely to use leftovers if they are willing to experiment and get creative, and that requires skill. With entire television channels and bookshelves devoted to cooking, it seems bizarre that fewer people are actually cooking in the home.

Are we hard-wired for convenience?

There are entire disciplines dedicated to understanding why we make purchasing decisions. I personally have a list of priorities when I buy food and these will shift in any given hour depending on my mood, financial circumstances, who's with me, and even the weather. With every choice there are costs and benefits and this is the foundation of understanding animal behaviours more generally. So, if you will indulge me in a little ecological perspective for a moment ...

Where, when and what an animal chooses to forage on requires a cost/benefit analysis. The nutritional benefits of that food are weighed against the likelihood of predation and the energetic cost of foraging (how hard they have to work for the food or how far they have to travel). The goal is obviously to maximise the energy while minimising the risk.

Animals use different strategies and adaptive behaviours to shift the cost/benefit ratio in their favour. Grazers are usually herd animals, minimising their risk of predation by being in a group. Social predators decrease the cost of foraging by hunting in packs and sharing. Many different animals, from insects to mammals, store food so that when food becomes less plentiful they don't have to travel as far to get it. Many animals use tools either to get at higher energy food or to process food that would otherwise be too energetically costly to try to access; chimpanzees, for example, use sticks to dig out termites, sea otters use a rock to crack open hard-shelled sea urchins, and crows will drop nuts onto hard surfaces from a height in order to crack them open.

Humans have mastered all of these strategies – we preserve and cache, we share, we use tools and we process food like no other creature on Earth. The biggest energetic cost we have is in getting up off the couch and driving to the supermarket. Our biggest risk is that we'll get food poisoning. On paper, the ready meal is the pinnacle of adaptive strategies for getting as much energy with as little effort as possible. Nothing else in the animal kingdom comes close to achieving this. If a bird were to maintain its juvenile plumage and receive regurgitated food from its parents for the rest of its life it still wouldn't come close to the benefits of the ready meal.

And yet, let me tell you a little story about two species of monkey. Howler monkeys and black-handed spider monkeys come from a common ancestor and occupy very similar habitats in that they spend all of their time in the forest canopy. These monkeys are similar in size and they are both vegetarians feeding on leaves and fruit. When fruit becomes less plentiful in the forest, howler monkeys just begin eating more leaves. They aren't as nutritious as fruit and because of the amount of fibre, they take a long time to digest, but the howler monkeys don't have to travel as far to get them. The spider monkeys, on the other hand, spend more energy travelling further in order to seek out what little fruit there is in the forest. In order to do this, these spider monkeys have evolved complex communication systems to tell members of their family when they've found fruit, and they have a wonderful memory that enables them to remember where fruits can be found in the forest in different seasons, so they don't waste energy looking. Meanwhile, the howlers are lying about all day digesting their incredibly high-fibre, low-energy food. As it turns out, the spider monkey's brain is twice as heavy as that of a howler monkey. There are always long-term consequences of short-term gains. So, I guess my question is, would you rather be a howler or a spider monkey?

CHAPTER EIGHT

Really Really Small Stuff

N ano is the new micro. Perhaps, like me, you are old enough to remember when everything was micro-something. There were microchips and micro-waves and the BBC Microcomputer. My library still had a microfiche and in spy films they were always after the microfilm. The first microfinance institutions were giving out microcredit to extend microloans. People were being micromanaged in their jobs and Bill Gates and Paul Allen started Microsoft. 'Micro' comes from the Greek word *mikrós*, which means small. It represents one-millionth or 10^{-6} and to give a sense of size, a typical bacterium is between 1 and 10 micrometres (also known as microns) in diameter. It's small. But then nano came along.

These days everything is nano. Now we have the iPod Nano, which you listen to as you do CrossFit wearing your Reebok Nano shoes. There is, of course, nanofinance and nanocomputing, which uses nanochips. Our phones use nanoSIM cards. Science fiction films are about nanorobots gone crazy and surgeons are using nanosurgery to change the genetic material of cancer cells. 'Nano' is also derived from a Greek word: *nanus*, which means dwarf. It represents one-billionth or 10^{-9}. The DNA helix is approximately 2.5 nanometres (nm) in diameter. It's really, really small.

I first became interested in nanoscience* when I wrote a story about how nano-sized iron could be used in the

---

* There are going to be a few nano-prefixed words used throughout this chapter, so let me clarify my definitions.

remediation of contaminated land. A couple of years later I was covering another story, this time about the fate and toxicity of nanoparticles in our water systems. I started to realise just how widespread they are in our society, from our underpants to our mobile phones, and from aeroplanes to skin cream. As I spoke with more and more researchers, I started to get the feeling that our understanding of how manufactured nanoparticles behave within living bodies and within our environment was not as far developed as our widespread application of them.

It wasn't until I was researching food fraud a few years ago that I first learned about the use of nanoparticles in food. Specifically, that nanosilver was used to coat asparagus in order to prevent microbial growth and increase its shelf life. I'm not going to lie to you, I had concerns.

So, when I received an invitation to be a speaker at a two-day summit in Braga, Portugal discussing nano-technology and food, I jumped at the opportunity to learn more.

The International Iberian Nanotechnology Laboratory (INL) couldn't stand in starker contrast to the ancient buildings of Braga's historic core only 20 minutes' walk away. The Cathedral of Braga, constructed, updated and modified (as cathedrals often are) between the eleventh and seventeenth centuries, is a gothic-meets-baroque marvel of arches and towers. The INL building is 20,000 square metres of state-of-the-art facilities dedicated to nanoscience. In the basement are labs that have been isolated from all of the vibrations, electromagnetic energy and other disturbances that we don't even notice, but that have

Nanoscience is the study of all things at the nanometre scale. Nanoparticles are nano-sized particles (less than 100 nanometres) – an example would be nano iron. Nanomaterials are objects or structures that contain nano-sized components. Nanotechnology is the application of nanoparticles or nanomaterials.

significant impacts on anything nano-sized. The basement is also where all of the equipment to visualise nanoparticles is housed, such as an atomic force microscope that has a tiny needle that runs over the surface of an object, much like the needle of a record player, drawing surfaces that are fractions of a nanometre in size – individual atoms! On the main floor of the facility are clean labs where everyone is running about (no, not running, walking ... *never* run in a clean lab – unless the experiment you've been working on has just metamorphosed into an autonomous killer, then *run*) in white suits with all hair, skin and other 'dirty' things covered. The facility has housing for visiting researchers, lots of garden space, lab space, offices, a cafeteria, a gym and, perhaps most impressive of all, an on-site child-care facility that's free for staff. Nanoscience is not a cheap endeavour.

Over two days I listened to talks and discussions about the role of nanoscience in the food industry. It became clear to me that, much like food processing as a whole, we shouldn't paint all of nanotechnology with the same pessimistic brush. I recognised the potential (as have many others) for nanotechnology to fall victim to the same public distrust as genetically modified organisms (GMOs), from which it could never recover. For this reason, I have given the application of nanotechnology to food its own, albeit brief, chapter.

The fundamental concept of nanoscience is that when things get down to the nanoscale they behave differently from how one would expect. Nanoparticles no longer seem to be governed by the rules of classical physics – for example, materials that barely interact with one another at a larger scale suddenly form strong bonds when they are cut down to the nanoscale. Nanoparticles also do things like self-assemble and stay suspended in gas or liquid; they might have a different colour, conduct electricity differently and have different boiling or melting points. Part of the reason for this is that the surface area-to-volume ratio at this scale is large. Huge, actually. A cube of iron measuring 1cm x 1cm x 1cm, for example, has a surface area of $6cm^2$ ($1cm^2$ x 6 sides). The same volume of nano iron (10nm x 10nm x 10nm) has a surface area of 6 million cm^2 (each nano-iron particle has $600nm^2$ surface area and $10^{14}nm^2$ in every square centimetre). With such a high surface area-to-volume ratio, it means that far more of the interesting reactive bits are out there exposed to the surrounding environment and other reactive species. This can be very useful in many different applications, including food.

Nano in nature

I have read many papers in food nanotechnology and they all have one thing in common: there are very few words that are less than a dozen letters long. So, when I came across Lorenzo Pastrana, Head of the Life Science Department at INL, who is a naturally gifted science communicator, I cornered him. We literally hid on a bench under the stairs of the lobby of the INL because he knew that if we went to his office we would be disturbed constantly. As it turned out, people even found us there eventually.

'We are surrounded by nanomaterials all the time,' Lorenzo points out to me. As one would expect, there are many things in nature that are nano-sized. They are in the atmosphere, they are in the sand on a beach and in the ash violently ejected from a volcano. Many viruses are measured in nanometres. The film of a soap bubble is just tens of nanometres thick. They have found nanotubes in the fangs of spiders. There are nano-sized structures on the surfaces of bird feathers that create the incredible structural colours in a peacock's tail. It's not all robots and computers.

'Nanoparticles also occur naturally in the food we eat,' continues Lorenzo. He tells me that not only are the micelles (the groupings of casein mentioned in Chapter 2) in milk on the nanoscale, but so are whey proteins and lactose. The nano-sized casein micelles carry nutrients within their matrix and keep them suspended within the milk, making them more accessible. To use a postal analogy, these little nano-sized envelopes are wrapping up important insoluble minerals, like calcium phosphate, into an easily digestible package for infant mammals, including humans. There is nothing concerning about that.

Nanoparticles aren't new. It's just that we got a lot more excited about them once we could see them, which happened in 1981 with the development of the scanning tunnelling microscope. And we are getting even more excited about them now because we can manufacture and manipulate them. And this is perhaps when people start to get a bit jittery about it all.

Nano in the supermarket

Nanotechnology has applications at every phase of food production, processing, distribution and storage. Food processing and production is the fifth-largest sector using nanotechnology (measured by patents filed), behind medical applications, construction materials, paper products and plastics. There are currently 1,200 companies globally with active R&D programmes in nanotechnology, with food applications estimated to be worth around £319 million (US$410 million).

Within the food we eat, there are a number of different sources of nanomaterials. There are naturally occurring nanoparticles, as previously mentioned. Then there are nanomaterials that have occurred incidentally as a result of processing methods: grinding materials, whipping cream and oil emulsions, for example, can produce nano-sized globules. There are nanomaterials that are added incidentally: for example, some additives will come in a range of particle sizes, including at the nanoscale, such as fortifying ingredients like calcium and iron. And then there are engineered nanomaterials, which have been designed and incorporated for a specific purpose.

There are two ways to manufacture nanoparticles within the food industry. The first is using what is known as a top-down approach, which is pretty much as it sounds – taking bigger things and cutting, milling or machining them down to the nanoscale. The second approach is known as bottom-up, and it takes advantage of the inherent behaviours of some nanoparticles to self-assemble or self-organise. This involves building or growing structures from smaller units. For example, Lorenzo and his colleagues are looking at whey proteins and their propensity to denature (unfold) and aggregate. They are using different methods to control this behaviour in order to form them into gels, fibres and tubes, which could then be used to control the release of functional ingredients and additives.

The challenge for manufacturing nanomaterials for the food industry, of course, is that we put our food through a lot. Nanoparticles, in general, tend to be unstable when heated. This is obviously a problem for most foods. So scientists must do their best to understand how the physicochemical properties of nanoparticles change through every phase of the manufacturing of the food, from interacting with other ingredients in the first phase through to what the consumer might do with the food once they bring it home (we don't always follow the instructions on the packaging). Then there is the complex journey of these particles once the food is eaten: exposure to enzymes, extreme pH and interactions with a diversity of gut flora. Food systems offer a level of complexity that make some engineering challenges seem incredibly straightforward, yet they are out there in our food products already.

At the food production end of things, nanotechnology has been used to improve fertiliser formulations by improving the efficiency with which plants take up specific nutrients – greater surface area increases uptake and so less fertiliser is needed. Nanomaterials can also be designed for slow release of nutrients or even pesticides. Nanoparticles have been incorporated into polythene films to help scatter the light entering polytunnels. This scattering effect does two things. First, it causes the light to hit the plants at all angles, increasing photosynthesis and improving growth, and ultimately increasing the yields of covered crop systems like strawberries. Second, by scattering the longer wavelengths of light it also reduces the heat generated in the polytunnels, stopping scorching during the hot summer months.

Nano-sized particles can be engineered for specific functions. For example, nanocellulose (usually fibres) forms a gel that is useful as a thickener or stabiliser, particularly in low-calorie foods. An Israeli startup company, DouxMatok, is using nanotechnology in the development of a new generation of sweetener that it hopes to have on the market by 2018. They use a nano-sized particle of silica or cellulose at the centre, both of which are commonly used food additives and both of which have oxygen atoms exposed on the surface. Molecules of sugar are then used to coat the nanoparticle, forming bonds with the exposed oxygen. The sugar molecules are perceived by the sweet receptors on the tongue and because they are bound to the core particle, it is thought that they don't disperse as quickly as an unbound sugar molecule would, prolonging the sweet flavour. The cellulose or silica has no caloric value or is thought not to be digested, so the same level of sweetness is delivered with far fewer calories. The company estimates that this new sugar could reduce the amount of sugar needed in food and beverage products by 25 to 55 per cent depending on the item. Scientists are also exploring ways to cut salt particles down to the nanoscale in order to take advantage of the surface area-to-volume ratio – far less salt would be needed to achieve the same flavour.

Remember those edible coatings on fresh-cut fruits and vegetables? Nanotechnology has a role here too. Nanoscale layers can be developed with lipids, proteins or polysaccharides as their foundation, which create a moisture barrier. Impregnate these layers with nanoparticles that have antimicrobial activity or other functions, and the coating takes on a new level of purpose. Instead of coating an apple in a layer of wax, as we do now, it might instead have a couple of nano-layers of lipid molecules. As you may have gathered from Chapter 4, I'm not a fan of edible coatings, but if they are going to use them then maybe it is better that they are thinner.

Nanoemulsions have droplet sizes that are at the nanoscale, which changes the flow of the liquid. Emulsions like mayonnaise or ice cream can have about a 15 per cent drop in fat content without using other additives, simply by changing the scale of the emulsion. There are products on the market now using this technology.

A considerable amount of research effort is currently going into nanoencapsulation, which is much like the natural micelles in milk delivering nutrients more efficiently, except scientists have decided what to deliver. Using naturally self-aggregating units, such as casein proteins, food scientists can engineer very small bubbles or coatings for different functions. These coatings might help to preserve an additive during processing or storage, or they might help to hide an unpleasant taste, for example, hiding the awful taste of iron by encapsulating it with casein. The coating might not be hiding its contents from taste buds, but rather the environment it's in – encapsulating a hydrophobic substance, for example, so that it disperses better in water. The capsules can be designed to disassemble with changes in pH, such as might happen during digestion, for example. And they can also be used to help improve the uptake of micronutrients, such as vitamins and minerals or even probiotics. Carotenoid pigments, which we get from fruits and vegetables, play an important role in cognitive function and eye health, but we have to be eating a whole lot of kale and spinach to get some of the really important ones. There are few of us who can manage such levels of herbivory. These pigments can be encapsulated in a lipid layer, which makes them easier to absorb, and therefore more available to the body. This could have significant health impacts, including reductions in age-related blindness. In short, nanoencapsulation can potentially reduce the amount of additives many of us loathe, or improve nutrition or perform any other number of functions within foods.

The largest research area, however, when it comes to the application of nanotechnology to food is in packaging. Nanomaterials are being used to improve the barrier properties of packaging: inert nanoparticles are generally dispersed within a polymeric material (the various types of plastic and polystyrene). These particles can change the permeability of the barrier to gases simply by getting in the way of these molecules, like some odd sort of pinball game. An oxygen molecule, for example, can't simply diffuse directly through the plastic; it has to weave its way through the nanoparticles, making its journey much longer, which means the packaging can be much thinner to achieve the same permeability. Nanoclay is currently used in the packaging of fruit juices and carbonated drinks for this purpose.

Composite plastics can incorporate nanomaterials to improve their strength and rigidity while reducing the amount of plastic required. This quality has been of particular interest in bio-based plastics where it is difficult to improve the strength of the packaging without affecting its biodegradability.

Nanotechnology can also play a far more active role in packaging systems. Many act as antimicrobials. Titanium oxide coatings in salad bags reduce the microbial contamination of fresh-cut lettuce, for example. The little absorbent pads underneath your packaged meat contain nanosilver, which helps to reduce the presence of aerobic bacteria. Other nanoparticles can actively scavenge oxygen, which might otherwise speed up the rotting process; plastic packages with palladium and zinc oxide dispersions or coatings are currently in use.

Finally, nanotechnology is helping to create intelligent packaging: containers that monitor what is going on inside and adapt to the situation or send up a red flare if something is wrong. Most of this is still in the research phase and not yet commercially viable, but it has promise. Scientists are currently exploring, for example, a system that releases a preservative into the food only if that food is put at risk – say it reaches a threshold temperature or age. This could help to decrease the amount of preservatives in food, deploying them only when they are needed. Platinum nanoparticles have the potential to act as a pH sensor to tell you if your bottle of wine has gone off. Copper nanoparticles coated in carbon have been used in a film layer as a sensor for moisture. When moisture inside the package increases, the film swells and the copper nanoparticles become less densely packed, causing a colour change in the thin indicator film. Similar indicators are being explored for detecting gaseous amines that are a sign that meat or fish has gone off, and ethylene gas that indicates fruit ripening. Smart packaging could help to significantly reduce food waste by avoiding all of the confusion around 'best before' and 'use by' dates. It could also help to reduce food poisoning, though I can also see the potential here for people to rely entirely on the packaging and less on their common sense.

While there's clearly a very practical role for nanotechnology in the food packaging sector to help reduce food waste and improve food safety, one also has to wonder what the industry will do for fun. How will they use nanomaterials to attract consumers into their products and enhance their eating and drinking experiences? There's already gimmicky packaging out there that incorporates things like temperature-sensitive inks. The Belgium-based beer brewers Anheuser Busch InBev feature thermostatic ink on their Oculto beer. There's a skull on the bottle and the eye sockets show leaves through them when it's at room temperature, but when it gets chilled a pair of eyes appear. Imagine what might be possible with nanotechnology?

Nano in our bodies

If understanding what nanoparticles will do in food is a challenge, understanding what they will do in the human body seems hopeless. The possibilities can be extremely simplified (I can't highlight the word 'extremely' enough here) as follows:

1. the nanoparticles ingested could be eliminated immediately, passing right through the alimentary canal unscathed, or

2. the nanoparticles could undergo a series of physical changes, as a result of enzymes or pH changes or any number of other things happening in the gut, where they form bigger aggregations (no longer nano-sized) or break into smaller particles or even into molecular components.

These components could then

a. try to pass through the mucous layer of the gut only to be spat back out again to be eliminated, or

b. pass through the mucous layer, where they could then

    i. get taken up by the epithelial cells and accumulate right there in those cells, or

    ii. get taken up by the epithelial cells and just pass through, or

    iii. pass between the epithelial cells.

If it's ii. or iii. they can then either

    i. enter the blood stream, or

    ii. enter the lymphatic system, or

    iii. get stored in nearby or distant tissues.

It's sort of like an odd choose-your-own-adventure story. And in the absence of evidence, this is roughly how our understanding of how these nanoparticles behave in our body has to be treated.

In order to understand what is happening to nanoparticles once they are ingested, a risk assessment can be carried out, which essentially explores all of these adventure-story endings. Is the material likely to undergo physical changes? If so, is it likely to be absorbed in the intestine? If so, where will it go from there and what are the risks? Unfortunately there's a paucity of information out there to help answer any of these questions. But scientists are working on it.

Nanosilica is perhaps one of the most studied nanomaterials out there, partly because it has been in wide use within the food industry for a very long time, although inadvertently. Silica is used as an anticaking agent and in Europe is listed as E551 on packaging. Because of the way it is manufactured, it comes in a range of sizes, including nanoparticles. The first step in understanding the health risks of ingesting nanosilica involves knowing how much of it is in the food. It turns out that this can vary dramatically between different processed foods. In instant asparagus soup 33 per cent of the silica is nano-sized, in coffee creamers it's 19 per cent and in pancake mixes it's less than 4 per cent. But it doesn't do just to look at the nanosilica content in the dry soup mix or the dry creamer; the food needs to be prepared as someone would consume it. For example, the coffee creamer has to be added to a cup of coffee and stirred around. When this is done, the amount of nanosilica more than doubles.

Nanosilica is clearly available in these foods, so what happens to it once it goes down the gullet? Scientists have used models that mimic human digestion. They are essentially big beakers that have the pH, enzymes and other qualities at levels which try to emulate the main components of the human digestive system: mouth, stomach, small intestine, colon. When researchers put silica (E551) through their fake gut, they found that nanosilica was present in the mouth. The enzymes in human saliva didn't seem to cause any changes in the particles at all. However, once the silica reached the stomach, no nanoparticles could be located. The researchers measured again once the food reached the fake intestine and found that the nanoparticles had reappeared again and in even higher amounts than were present in the mouth. The researchers concluded that the low pH and high electrolyte concentration in the stomach caused the nanoparticles to agglomerate, forming larger groupings. Based on a fake digestive system, at least, it would seem as though nanosilica is reaching the intestinal epithelium.

As to the fate of the nanosilica once it reaches the intestine, it is still mostly unknown. Experiments on rats and mice have shown that nanosilica causes liver toxicity, though the effect wasn't there when the silica particles were micron-sized (that is, 1,000 times bigger than a nanoparticle), suggesting the toxicity is a result of the size rather than the type of particle.

Most governments have set a safe dose limit for silica, as with all food additives. However, at the nanoscale things change. Perhaps more of the silica is absorbed by the intestine, or perhaps less. If more is absorbed, then these particles have a much greater surface area ratio, which might make them far more biologically active within the body, which skews dosage values.

The fate of nanoparticles in the human body is poorly understood and while it is obviously an area of keen interest, there are severe limitations. First of all, trying to find these particles in the human body would be like dropping a Tic-tac into a fast-flowing river and trying to find it again. Did it drop straight to the bottom or did it get carried downstream? If so, how far? Maybe it got stuck in a back eddy. Did it dissolve? Or perhaps the outer colour dissolved so now it looks different. Maybe a fish ate it. Add to this the challenge that not many human subjects are terribly keen on volunteering for experiments where they are asked to consume something that may or may not have a toxic effect. As a result we are left to play with beakers of gastric juices and to experiment on animals.

So here we are faced with a technology that offers such enormous potential, which could without a doubt revolutionise food systems and be the biggest change in processing since refrigeration. And yet we are a little scared to let it loose because there remains so much uncertainty. What do we do?

I put the question to Lorenzo, trying to understand how he rationalises it all in his own research. 'My team and I focus on the use of nanoparticles that are already in existence in the human diet,' he explains. 'We of course look closely at how these particles behave, but by looking at foods that already have nanostructures, we feel we are starting in the right place.'

Nano in our perceptions

Public knowledge of nanomaterials is somewhat limited, particularly when it comes to food. The British consumer watchdog organisation Which? conducted a survey in 2008 and found that 61 per cent of shoppers have never heard of nanotechnology. Before I started writing about food, if someone was to ask me about nanomaterials, I would probably have drawn examples from the tech sector – computers and smartphones – or perhaps nanocomposite materials that prevent ice build-up or disperse lightning strikes on aircraft. Food applications, however, do not immediately spring to mind.

When people know little about a technology, then their perceived risk of that technology will depend on who's using it and whether they trust them. Consumers generally don't have a lot of trust in the food and beverage industry, despite the fact that any of us who don't grow 100 per cent of our own food depend on them (it's a bit of an odd relationship, really). As new nanotechnology is revealed, its acceptance among the public will be likely to depend on who delivers it. For example, a technology developed by a university is likely to be perceived differently from one developed by a multinational corporation.

Consumers also seem more willing to accept nanotechnology when it comes to packaging rather than ingredients. There is less perceived risk when you're not eating them, and yet we eat naturally occurring nanoparticles all the time.

If consumers see that there are multiple benefits of nanotechnology, then their perceived risk goes down. If a new form of packaging could improve the safety and shelf life of a food product while making the packaging more environmentally friendly by reducing the plastic needed, it would probably be welcomed with open arms. However, if there was some uncertainty as to whether the nanoparticles in that packaging migrated into the food or if it was unclear how they dispersed into the environment at the end of the package's life cycle, then uneasiness increases again. Consumers want to know that the people delivering the nanotechnology can control the nanotechnology.

For me personally, I think I would be more accepting of nanotechnology that is in line with my ethics. If it helps the environment and reduces food waste, then bring it on. If it's a new fancy nanoencapsulated particle that delivers an intense flavouring right to my taste buds, well, I think I can probably survive without it.

Add to all of this the fact that regulation around nanotechnology is still unclear – not only for the consumers, but I fear for the industry as well. As of 2013, all ingredients present as nanomaterials have to be labelled with 'nano' in brackets. This is yet another layer of information on labelling that consumers can't do much with. Don't get me wrong, I think it should be on there, but unless you know the potential effect of a nanomaterial when you eat it, how do you really know how to feel about it?

There needs to be considerably more communication groundwork put in by government and industry around nanotechnology in our food. And I don't mean just lip service, I mean some good honest accessible information. This will build up trust in this tenuous relationship that we seem to have between consumer and industry. In the absence of information, we can let our imaginations get the better of us. This is technology that can cross into our cells and affect the machinery within. But we, as consumers, shouldn't always jump to the worst conclusion the second there's a gap in the science and assume that we're all being slowly poisoned to death. Take a breath and put some nanoparticles into the perspective of the risk associated with driving your car to work every day, or of UV rays mutating your DNA, or crossing the street while texting.

I personally feel that there is considerable potential for nanotechnology to address some of the challenges facing the food industry and society more broadly. But I'm also cautious of unleashing such powerful technology when there is so much uncertainty. I'm pragmatic enough to accept that nanotechnology has been and will continue to be introduced while there are still enormous gaps in our knowledge. There are obviously risks to that. But I think there is far more risk in fear-based campaigns tainting an entire area of research as a result of those gaps. I want to make sure that there is funding to support scientists like Lorenzo, who approaches nanotechnology with equal parts fervour and caution, and who is motivated by making the world a better place rather than by investors and bottom lines. As consumers we must be discerning. Like it or not, nanotechnology is upon us. In order to be discerning consumers, we will need scientists like Lorenzo to help communicate the benefits and risks associated with these technologies and we will need journalists interested in facts rather than fear-mongering. It is only then that we consumers will be properly equipped to support those technologies that have clear benefits and minimal risks, while opposing those where the risks appear to outweigh any benefits.

The Future of Food Processing

When I was conducting my research for this book, I came across a paper written in 2010 that caught my attention. In line with what I had been thinking, the authors couched food processing in the context of human evolution, and instead of talking about the physical properties of food, their viscosity, consistency and mouth-feel, wrote about sustainability and reducing food loss. It was a welcome change from all the other food science papers I had been reading as part of my research. I contacted the Institute of Food Technologists and was put in touch with the first author on the paper, John Floros, Dean of the College of Agriculture at Kansas State University and Director of K-State Research and Extension. John is also the former President and a current Fellow of the Institute of Food Technologists. I called him up.

Like anyone associated with food processing, John seemed to be weary of me. He seemed poised to defend processed food to yet another journalist. Or perhaps it's just because he had an unpleasant cold. We both did. It was February and the two of us were sniffing and coughing away. After a few minutes of me babbling on about my desire to present a balanced view of food processing, John started in with the big picture. 'Fifty years ago we had half the human population we have today,' he said. 'How did we manage to feed all of these people during this time of explosive growth? Looking forward, how are we going to feed the expanding population of the future? How are we going to provide safe, high-quality, wholesome foods to such a diverse population in a sustainable way?'

This is something I think about a lot and it worries me deeply, but I resisted the temptation to yell into the phone 'I have no ****ing clue John, help me figure it out!'

He continued: 'Agriculture is only one part of the solution, because we can't get more land, or water, or energy necessarily. We can improve efficiency and yields with genetics and cultivation practices, which we are working on, but we are wasting or losing one-third of the food we grow already. This brings us right back to food processing and technologies that give us the ability to preserve that initial product. Food science, and particularly processing, has a huge role to play in feeding the world as sustainably as possible by helping to decrease food loss and waste, and reduce the environmental impact of the food system.'

This is not the context in which the media discuss food processing. It's refreshing. It's hopeful.

I asked John what the pivotal advancements in food science have been most recently and he started to go into packaging and new methods of heating to reduce microbial contamination, but then he stopped. Under the weight of responsibility for training future food scientists, John realised that he needed to change the direction of the conversation from a stock-standard list of processing methods to something far more pivotal to our discussion.

'As a society, we have stopped investing in better processing methodologies partly because processing has become a dirty word,' he said. 'Social pressures have driven governments and funding agencies to withdraw financial support in these areas. There is still some money to do research in food safety because that remains a priority, but the support to drive innovation in the broader food processing area is simply not there. As a result, we have delegated all of that type of research to private companies who are responsible to their investors and motivated to improve their bottom line. The US is not alone in this.'

This hadn't occurred to me before. Because of our fickle relationship with processed food we have inadvertently created the situation we most wanted to avoid. We have tied the hands of the scientists who are most interested in taking processed food in the direction we would like to see it go – healthful and sustainable – and we have enabled multinational corporations that are keen to get their marketing departments to sell us what they most want to produce.

So our task now, as consumers, is to undo this wrong. We must be critical of what we hear about processed food and sort the fact from the fiction. We have an opportunity to use our purchasing power to make our priorities clear, so that we might help to bring about the food future we want. It's unlikely that we all want the same future, but it's probably safe to say that there are some scenarios we'd all like to avoid.

To start thinking about the future of food, it might first be useful to look at some emerging trends that are likely to continue to be impactful in most future food scenarios.

There is a continuing trend towards more 'natural' ingredients. Consumers don't want a list of complex chemical names on the label that mean nothing to them, they want things that they can recognise. In response, food manufacturers want to clean up their food labels. In order to achieve this, existing and potentially new natural sources of compounds will need to be identified that have the same functionality as the synthesised powder they are replacing. Then, economical methods for extracting, isolating and purifying these compounds will need to be developed. In the end, you will probably end up with essentially the same powder in a beaker, but with potentially greater environmental consequences. This isn't always the case, obviously, but the definitions around 'natural' ingredients remain hazy for a reason. Under all future scenarios, it is safe to say that there will be environmental uncertainty and resource limitations. Not only do 'natural' ingredients potentially put greater strain on resources, they are also more prone to environmental uncertainty. This can mean that their supply chains are unreliable and prices can be volatile, a scenario that also presents opportunities for food crime.

In stark contrast to the previous trend, it is likely that flavours in the near future are going to get extreme. It is predicted that new, previously unheard-of combinations of flavours are going to make their way into our food. According to *New Food Magazine*, breakfast food seems to be the meal most in need of a flavour overhaul; brown sugar and cinnamon are out of the porridge bowl and mango and cardamom are in. Personally, I welcome a bit of a breakfast shake-up, quite frankly. But I get cautious around discussions of flavours – some cutting-edge chef combining new flavours in the kitchen is one thing, but a team of analytical chemists in a lab is another. In Mark Schatzker's book *The Dorito Effect*, he thoroughly explores the world of flavour. It's an excellent and thought-provoking read. Flavour factories are developing new and exciting flavours all the time, which manufacturers incorporate into new product lines of snacks, beverages and sometimes even real food. My personal issue with manufactured flavours is that these aren't just regular flavours – they are super-flavours, designed to trigger specific sensory pathways that make them that much more obvious and vivid than the real thing. This alters our perception of what real food tastes like and, just like sugar, changes our frame of reference for all other foods. I think that under different future food scenarios, flavours could play either an important or a very diminished role. If circumstances shift considerably and we are doing whatever we can to meet our basic nutritional requirements, I think innovative new flavour combinations are probably going to be less important. If, however, we have gone so far as to embrace technologies that have removed all the aromas and flavours of food and reduced it to perfectly portioned combinations of macro- and micro-nutrients, I might start to embrace flavour of any kind.

Convenience is going to continue to trend while 'busy' is in fashion. And there's no sign of that letting up anytime soon. My greatest concern with convenience, as I have said, but will reiterate because I think it so critical, is the loss of a critical skill within our culture. Jamie Oliver* has been quoted as saying that children know more about drugs than they do about courgettes. Most of our children live in cities and have already lost a connection with their food. If parents aren't cooking with their children then they are widening that disconnect. If parents aren't preparing food in the home then their children can't even acquire food knowledge as passive observers. They have zero chance of acquiring the ability to feed themselves through their domestic life. They will know about stranger danger and e-safety, they can get to the nth level of some video game, but they can't feed themselves. I know first-hand how difficult it can be sometimes to include children in the process of cooking –first there's the whingeing, then there's the mishaps, followed by the inevitable mess. But who else is going to teach them these skills? As I've already dedicated an entire chapter to the topic of convenience, I won't dwell on it any more, except to say this. Stop. Cook. Eat. Enjoy. Do it with your partner, do it with your kids, do it as often as you can. There may be far more benefits to doing this than just a home-cooked meal.

To counter convenience, there is also a trend towards local, artisanal and slowly prepared food. Many consumers are looking to have more of a story associated with their food. They want to get to know their baker and talk to them about their craft. Consumers want the experience of tasting the unique flavours of a cheese ripened in a Welsh

---

* I realise I have now mentioned Jamie a few times, so I must publicly declare my admiration for him and what he and his foundation are trying to achieve in terms of food education – something I feel incredibly passionate about. Jamie, if you read this, I want to work with you!

mine. As a result of this, supermarkets are introducing brands with a local and artisanal feel to them. Beware of this, however. Tesco's launch of its own-brand meat and produce labelled with fictional British-sounding farm names received considerable criticism when it was launched in 2016; lettuce from 'Nightingale Farms' and fruit from 'Rosedene Farms' sounds like an attempt to source local produce, but it is simply a branding campaign. As we have seen in previous chapters, enzymes and additives can be used to give cheese a unique flavour or bread a sourdough taste. I personally love food with a good story, so I'm happy to see this trend evolving. As consumers, we just need to make sure we're getting the whole story.

Another trend that is on the cusp of emerging is personalised nutrition. The first step to this is already taking off: accessible DNA tests. Over here in the UK, the focus of these services has been around understanding your genealogy. The advertisements show thrilled customer after thrilled customer as they discover that they are 1/18th Icelandic and a quarter Neanderthal (I don't think they include that test, but that would be interesting). However, some of these tests are also starting to identify markers that may indicate a predisposition to certain diseases. This might trigger lifestyle and diet changes that lower your risk of acquiring those diseases. However, the next step in this service is to provide customers with information about how their unique DNA dictates how they metabolise and respond to certain foods. The partner to this is then of course to design foods that are optimal for the customer, most likely in the form of a pill or shake. If this sounds unappealing, consider how many people are currently substituting one or more meals a day with a powder-based substance that bears no resemblance to food, simply for the promise of weight loss. I can think of two or three people in my immediate circle. How many more people would be willing to eat like this if the promise was a long and disease-free life? While I find this idea intriguing, I can't help but think that this would require the services of chemists rather than food scientists. If we don't trust that it's beef in our lasagne, how could we ever trust that the powder in a little capsule is what it's supposed to be? Yikes, it's an authenticity nightmare!

The final trend I want to mention (once again in stark contrast to the previous trend) is eating insects. With a rising demand for protein and increased focus on sustainability, insects are the obvious choice. Two billion people in the world are already eating insects as a regular part of their diet. It is expected that by 2024, the edible-insect sector will be worth approximately £601 million (US$773 million). So, prepare yourself. This is going to open up all sorts of interesting opportunities and potentially new innovations for processing.

But now on to broader thinking about our food futures. The Institute for The Future (IFTF) offers four different scenarios for how our future might look when it comes to food. The first, called 'Growth', is very much a business-as-usual scenario, where consumers of the future will be able to access any food that they want at all times of year for reasonable prices. Food prices will drop thanks to breakthroughs in technologies that provide cheap alternative fuel sources, which bring down the costs of producing, manufacturing and transporting food. Technology will continue to be applied to challenges such as drought and disease resilience in crops as well as water access. Cheap food would mean that retailers and consumers, in particular, would at best only make incremental improvements in reducing food waste. And yet these technologies will not have been able to resolve inequalities in the distribution of food. Under this scenario, the IFTF suggests that diet-related health risks will continue to grow and overshadow all other sources of illness, remaining the dominant cause of mortality. I'm not sure how any healthcare system could survive under this burden; it would no doubt push countries into new crises.

The second scenario is called 'Constraint' and involves a significant global crisis that triggers swift and significant change. The IFTF proposes that a major global food-poisoning outbreak causes a complete collapse in consumer trust. In response, countries tighten their borders to food trade. Things would be bleak for a while – choice would be limited. Countries that devote vast expanses of their agricultural land to a single crop, such as wheat or corn, would have to diversify. Communities could respond by converting urban spaces into areas for food production – perhaps as guerrilla gardeners initially, but then with the support of government as it became clear that urban food production could help to meet demands. Research helps to advance urban production. There are international treaties that prevent the production of resource-demanding food and place high tariffs on the movement of food. Despite the obvious hardships initially, there are a lot of things about this scenario that I like. It speaks to me of a return to local economies, with local artisan products. I love the idea of urban food production and building in food resilience at a community level. But on the other hand, if a region was to experience a crisis, say an earthquake, would we have lost the infrastructure for shipping food to these places quickly and without waste? Also, the idea of an event so large as to trigger such a response is terrifying to me, particularly at a time when terrorism threats are high and food systems are vulnerable. However, it might also take much smaller events to change the flow of food in the world. It's estimated that 97 per cent of the UK's food exports are going to be affected by our leaving the EU. The current US President Donald Trump is most certainly tightening borders; so far he's targeted people, but who knows what executive orders are in the pipeline. This scenario is not dissimilar from what people experienced during the constraints of both World Wars. There was less choice and people were forced to buckle down and not waste food. They began to 'dig for victory', 'can all you can' and 'grow your own'. People are so resilient when faced with extreme events and in fact such hardship can foster innovation – Napoleon's sugar beets being an excellent example.

In the third scenario, aptly named 'Collapse', agriculture collapses as a result of the loss of pollinators. This catastrophe seems set in the context of widescale environmental uncertainty globally. Many countries respond by putting in place rationing schemes and hoarding whatever resources they have.* Nutritional deficiencies and their associated diseases begin to rear their ugly heads. Food manufacturers struggle to find ways to replace key ingredients in their products, but most fail. Countries go to war over resources. This is an alarming scenario to say the least. In this scenario, food scientists would no doubt be concentrating on functional foods to fill the dietary void created in such a catastrophe. Protein bars and supplements would no doubt be the main menu items. As with all future thinking, this scenario works in challenges that we are already facing. More than half the wild bee species in the US were lost during the twentieth century. Luckily for us, nature is incredibly resilient. Where one species declines, another fills its niche, pollinating that crop. But just how resilient can our pollinators be? Three-quarters of global food crops require pollinators, and wild pollinators are doing twice the work of honey bees. I have suddenly been filled with dread at the thought of millions of miniature drones trying to perform such a function – oh, the horror!

In the IFTF's fourth and final scenario, called 'Transformation', they paint a very sci-fi world of cultured meat and 3D-printed food. Yet, in the six years since the IFTF's report was published, both of these have become a reality. 3D food printing began with the idea that it might

---

* Heck, maybe the US would finally crack open that cave of processed American cheese!

be a viable option for long-haul space travel. It has so far found greater utility on Earth. 3D food printers work on the same principle as any deposition 3D printer, except that instead of building up layers of plastic or metal, they're pumping out food mixtures with new levels of precision. They are currently printing everything from pizzas to chocolate. Obviously the source material for pizza is going to be quite different from chocolate. And this is where things break down a little at the moment. Some restaurants are already using a 3D printer to print things like cake decorations or intricate chocolate designs to adorn a top-notch dessert. In this case, icing or chocolate is put into the printer and it just pumps out perfectly uniform designs. However, a pizza would require different input mediums for the crust, sauce and cheese and these would need to be modified substantially in order to make them printer-ready. So, the idea of having a printer at home that could just whip up something on a whim is still out of reach.

In the 'Transformation' scenario, university dorm rooms are equipped with 3D printers and while not every household has one, community organisations might invest in communal services. There is an entire industry that develops around this, providing printer-friendly meals for download. 3D printed food is considered the 'fresh' alternative to the ready meal. The IFTF report rightly points out, however, that not everyone would be on board for such a future. There would be advocates of traditional food preparation calling themselves Authentic Eaters (I would definitely be part of this group). Some food manufacturers would build their brand around 'Made by Humans' labelling and certification programmes. The main themes of this scenario are customisation and convenience. And while some aspects might feel far-fetched, it is definitely a future that takes into account our willingness to embrace technology and our love for convenience.

Personalised nutrition would definitely be happy in this scenario, but there might also be further customisation of food and beverages for specific activities – a pre-'big night out' shake, for example, or a post-exercise smoothie, or a pre-bed shot. This scenario also hints at a divergence in societal values between the cookers and the non-cookers. This is a fascinating concept and we are already living aspects of this scenario; the technology just isn't cheap enough yet.

I think that we're on the verge of some very funky things happening with food. Students at MIT have just produced 4D shape-shifting pasta that comes out of the package flat and then contorts into different shapes when it is boiled. The motivation behind the idea was that pasta takes up less space when it's flat, so less packaging would be required. Yet, it's the shapes that make pasta fun. After 3D printing the pasta, the students printed different patterns of cellulose over the top layer of the pasta. The cellulose caused the pasta to curl into different shapes in the boiling water, depending on its pattern. This may seem entirely far-fetched and crazy today, but it's uncertain what challenges lie ahead. If we don't tinker a little now, we won't have the breadth of knowledge we may need to overcome these challenges.

Food processing has a role to play within all of these potential scenarios. But efforts must be directed towards the challenges that face humanity and the planet, rather than the profit targets of corporations. As much as I would love to end this book with a list of dos and don'ts when it comes to processed food, this is simply an impossibility. Each of us has our own priorities that drive our purchasing decisions – we may be motivated by cost, environmental footprint, animal welfare or time-saving options. Luckily, there is a plethora of choices available to meet whatever motivates us in the moment, and processing has been largely responsible for providing us with this choice. What

I hope I have managed to do for you, though, is present a balanced picture of food processing and challenge some of the common misconceptions. Armed with such information, we can be more rational consumers, capable of discerning what is acceptable to us in terms of processing and what is not. In this way we are better equipped to shape the food future each of us desires.

Selected References

Introduction

Floros, J.D., Newsome, R., Fisher, W., Barbosa-Cánovas, G.V., Chen, H., Dunne, C.P., German, J.B., Hall, R.L., Heldman, D.R., Karwe, M.V., Knabel, S.J., Labuza, T.P., Lund, D.B., Newell-McGloughlin, M., Robinson, J.L., Sebranek, J.G., Shewfelt, R.L., Tracy, W.F., Weaver, C.M. & Ziegler, G.R. 2010. Feeding the world today and tomorrow: the importance of food science and technology. *Comprehensive Reviews in Food Science and Food Safety* 9: 572–99.

Gelata, B. 2011. A world of hunger amid plenty. http://www.ifrc.org/en/news-and-media/opinions-and-positions/opinion-pieces/2011/a-world-of-hunger-amid-plenty/

Chapter 1

Casini, L., Contini, C., Marone, E. & Romano, C. 2013. Food habits. Changes among young Italians in the last 10 years. *Appetite* 68: 21–9.

Decker, E.A., Elias, R.J. & McClements, D.J. (eds). 2010. *Oxidation in Foods and Beverages and Antioxidant Applications: Management in Different Industry Sectors*. Elsevier.

Eurobarometer. 2010. *Food-related risks*. Special Eurobarometer 354.

Floros, J.D. et al. 2010. Feeding the world today and tomorrow.

Gowlett, J.A.J. & Wrangham, R.W. 2013. Earliest fire in Africa: towards the convergence of archaeological evidence and the cooking hypothesis. *Azania: Archaeological Research in Africa* 48: 5–30.

Koebnick, C., Strassner, C., Hoffman, I. & Leitzmann, C. 1999. Consequences of a long-term raw food diet on body weight and menstruation: results of a questionnaire survey. *Nutrition & Metabolism* 43: 69–79.

Lieberman, D.E., Krovitz, G.E., Yates, F.W., Devlin, M. & St Claire, M. 2004. Effects of food processing on masticatory strain and craniofacial growth in a retrognathic face. *Journal of Human Evolution* 46: 655–77.

Organ, C., Nunn, C.L., Machanda, Z. & Wrangham, R.W. 2011. Phylogenetic rate shifts in feeding time during the evolution of *Homo*. *Proceedings of the National Academy of Sciences* 108: 14555–9.

Plazzotta, S., Manzocco, L. & Nicoli, M.C. 2017. Fruit and vegetable waste management and the challenge of fresh-cut salad. *Trends in Food Science & Technology* 63: 51–9.

Ramey, V.A. 2009. Time spent in home production in the Twentieth-Century United States: new estimates from old data. *The Journal of Economic History* 69: 1–47.

UN Food and Agriculture Organization. 2011. *Global food losses and food waste – Extent, causes and prevention.* Rome.

Varrela, J. 1992. Dimensional variation of craniofacial structures in relation to changing masticatory-functional demands. *European Journal of Orthodontics* 14:31–6.

Wansink, B. 2004. Consumer reactions to food safety crises. *Advances in Food and Nutrition Research* 48: 103–50.

Welch, R.W. & Mitchell, P.C. 2000. Food processing: a century of change. *British Medical Bulletin* 56: 1–17.

Zink, K.D. & Lieberman, D.E. 2016. Impact of meat and Lower Palaeolithic food processing techniques on chewing in humans. *Nature* 531: 501–3.

Chapter 2

Azarnia, S., Robert, N. & Lee, B. 2006. Biotechnological methods to accelerate cheddar cheese ripening. *Critical Reviews in Biotechnology* 26: 121–43.

Bachman, H.-P. 2001. Cheese analogues: a review. *International Dairy Journal* 11: 505–15.

Blundel, R. & Tregear, A. 2006. From artisans to 'factories': the interpenetration of craft and industry in English cheese-making, c.1650–1950. *Enterprise & Society* 7: 705–39.

Burger, J., Kirchner, M., Bramanti, B., Haak, W. & Thomas, M.G. 2007. Absence of the lactase-persistence-associated allele in early Neolithic Europeans. *Proceedings of the National Academy of Sciences* 104: 3736–41.

Copley, M.S., Berstan, R., Dudd, S.N., Docherty, G., Mukherjee, A.J., Straker, V., Payne, S. & Evershed, R.P. 2003. Direct chemical evidence for widespread dairying in prehistoric Britain. *Proceedings of the National Academy of Sciences* 100: 1524–9.

Dudd, S.N. & Evershed, R.P. 1998. Direct demonstration of milk as an element of archaeological economies. *Science* 282: 1478–81.

Dunne, J., Evershed, R.P., Salque, M., Cramp, L., Bruni, S., Ryan, K., Biagetti, S. & di Lernia, S. 2012. First dairying in green Saharan Africa in the fifth millennium BC. *Nature* 486: 390–4.

Evershed, R.P. 2008. Organic residue analysis in archaeology: the archaeological biomarker revolution. *Archaeometry* 50: 895–924.

Evershed, R.P. 2008. Earliest date for milk use in the Near East and southeastern Europe linked to cattle herding. *Nature* 455: 528–31.

Fox, P.F., Guinee, T.P., Cogan, T.M. & McSweeney, P. L. H. *Fundamentals of Cheese Science* (2nd edition). Springer New York, 2017.

Gerbault, P. 2011. Evolution of lactase persistence: an example of human niche construction. *Philosophical Transactions of the Royal Society London B Biological Sciences* 366: 863–77.

Guldfeldt, L.U., Sørensen, K. I., Strøman, P., Behrndt, H., Williams, D. & Johansen, E. 2001. Effect of starter cultures with a genetically modified peptidolytic or lytic system on Cheddar cheese ripening. *International Dairy Journal* 11: 373–82.

Itan, Y., Powell, A., Beaumont, M.A., Burger, J. & Thomas, M.G. 2009. The origins of lactase persistence in Europe. *PLOS Computational Biology* 5: e1000491.

Law, B.A. 2011. Controlled and accelerated cheese ripening: the research base for new technologies. *International Dairy Journal* 11: 383–98.

Lindsay, R.C., Hargett, S.M. & Graf, T.F. 1980. Preference evaluation of foods prepared with imitation cheeses. *Food Product Development* 14: 30.

Mistry, V.V. 2001. Low fat cheese technology. *International Dairy Journal* 11: 413–22.

Mohamed, A.G. 2015. Low-fat cheese: A modern demand. *International Journal of Dairy Science* 10: 249–65.

Moskowitz, G.J. & Noelck, S.S. 1987. Enzyme-modified cheese technology. *Journal of Dairy Science* 70: 1761–9.

Padberg, D.I., Knutson, R.D. & Jafri, S.H.A. 1993. Retail food pricing: horizontal and vertical determinants. *Journal of Food Distribution Research* 24: 48–59.

Salque, M., Bogucki, P.I., Pyzel, J., Sobkowiak-Tabaka, I., Grygiel, R., Szmyt, M. & Evershed, R.P. 2013. Earliest evidence for cheese making in the sixth millennium BC in northern Europe. *Nature* 493: 522–5.

San Martín-González, M.F., Welti-Chanes, J. & Barbosa-Cánovas, G.V. 2006. Cheese manufacture assisted by high pressure. *Food Reviews International* 22: 275–89.

Chapter 3

Anon. 1848. Adulteration of Bread. *Scientific American* 4: 38.

Biagi, F., Klersy, C., Balduzzi, D. & Corazza, G.R. 2010. Are we not over-estimating the prevalence of coeliac disease in the general population? *Annals of Medicine* 42: 557-61.

Bollet, A.J. 1992. Politics and pellagra: the epidemic of pellagra in the US in the early twentieth century. *The Yale Journal of Biology and Medicine* 65: 211-21.

Campbell, J., Hauser, M. & Hill, S. *Nutritional Characteristics of Organic, Freshly Stone-ground, Sourdough & Conventional Breads*. Ecological Agriculture Projects – Publication 35. http://eap.mcgill.ca/publications/EAP35.htm

Cauvain, S. 2015. *Technology of Breadmaking* (3rd edn). Springer, Switzerland.

Collado, M.C., Donat, E., Ribes-Koninckx,C., Calabuig, M. & Sanz, Y. 2009. Specific duodenal and faecal bacterial groups associated with paediatric coeliac disease. *Journal of Clinical Pathology* 62: 264–9.

Costabile, A., Santarelli, S., Claus, S.P., Sanderson, J., Hudspith, B.N., Brostoff, J., Ward, J.L., Lovegrove, A., Shewry, P.R., Jones, H.E., Whitley, A.M., Gibson, G.R. 2014. Effect of breadmaking process on *in vitro* gut microbiota parameters in irritable bowel syndrome. *PLoS ONE* 9: e111225.

Drummond, J.C. 1940. The Nation's larder in wartime: food in relation to health in Great Britain. *The British Medical Journal* 1: 941–3.

Dunne, J., Mercuri, A.M., Evershed, R.P., Bruni, S. & di Lernia, S. 2016. Earliest direct evidence of plant processing in prehistoric Saharan pottery. *Nature Plants* 3: 16194.

Editor. 1878. The Assize of Bread. In *Archaeologia Cantiana* Volume 12. Kent Archaeological Society: Proceedings, 1877–8, p. 321.

Fasano, A., Berti, I., Gerarduzzi, T., Not, T., Colletti, R.B., Drago, S., Elitsur, Y., Green, P.H.R., Guandalini, S., Hill, I.D., Pietzak, M., Ventura, A., Thorpe, M., Kryszak, D., Fornaroli, F., Wasserman, S.S., Murray, J.A. & Horvath, K. 2003. Prevalence of celiac disease in at-risk and not-at-risk groups in the US. *Archives of Internal Medicine* 163: 286–92.

Fasano, A. & Catassi, C. 2012. Celiac disease. *New England Journal of Medicine* 367: 2419–26.

Gallone, B., Steensels, J., Prahl, T., Soriaga, L., Saels, V., Herrera-Malaver, B., Merlevede, A., Roncoroni, M., Voordeckers,K., Miraglia, L., Teiling, C., Steffy, B., Taylor, M., Schwartz, A., Richardson, T., White, C., Baele, G., Maere, S. & Verstrepen, K. J. 2016. Domestication and divergence of *Saccharomyces cerevisiae* beer yeasts. *Cell* 166: 1397–410.

Haaland, R. 2007. Porridge and pot, bread and oven: food ways and symbolism in Africa and the Near East from the Neolithic to the present. *Cambridge Archaeological Journal* 17: 165–82.

Lück, E. 1990. Food applications of sorbic acid and its salts. *Food Additives & Contaminants* 7: 711–15.

Mahadov, S. & Green, P.H.R. 2011. Celiac disease. *Gastroenterology & Hepatology* 7: 554–6.

Montonen, J., Boeing, H., Fritsche, A., Schleicher, E., Hans-Georg, J., Schulze, M. B., Steffen, A. & Pischon, T. 2013. Consumption of red meat and whole-grain bread in relation to biomarkers of obesity, inflammation, glucose metabolism and oxidative stress. *European Journal of Nutrition* 52: 337–45.

Mortimer, R.K. 2000. Evolution and variation of the yeast (*Saccharomyces*) genome. *Genome Research* 10: 403–9.

Mostad, I.L., Langaas, M. & Grill, V. 2014. Central obesity is associated with lower intake of whole-grain bread and less frequent breakfast and lunch: results from the HUNT study, an adult all-population survey. *Applied Physiology, Nutrition, and Metabolism* 39: 819–28.

Samsel, A. & Seneff, S. 2013. Glyphosate, pathways to modern diseases II: Celiac sprue and gluten intolerance. *Interdisciplinary toxicology* 6: 159–84.

Samuel, D. 1989. Their staff of life: initial investigation on ancient Egyptian bread baking. In Kemp, B.J. (ed.). *Amarna Reports V.* London: Egypt Exploration Society 253–90.

Samuel, D. 1996. Investigation of ancient Egyptian baking and brewing methods by correlative microscopy. *Science* 273: 488–90.

Shelly, C.E. 1924. Millstone flour and national nutrition. *British Medical Journal* 1: 1075.

Tuhumury, H.C.D., Small, D.M. & Day, L. 2014. The effect of sodium chloride on gluten network formation and rheology. *Journal of Cereal Science* 60(1): 229–37.

Wieser, H. 2007. Chemistry of gluten proteins. *Food Microbiology* 24: 115–19.

Wilson, B. 2008. *Swindled: The Dark History of Food Fraud, from Poisoned Candy to Counterfeit Coffee.* Princeton University Press, New Jersey.

Chapter 4

Baselice, A., Colantuoni, F. Lass, D.A., Nardone, G. & Stasi, A. 2017. Trends in EU consumers' attitude towards fresh-cut fruit and vegetables. *Food Quality and Preference* 59: 87–96.

Ekwaru, J.P., Ohinmaa, A., Loehr, S., Setayeshgar, S., Xuan Thanh, N. & Veugelers, P.J. 2017. The economic burden of inadequate consumption of vegetables and fruit in Canada. *Public Health Nutrition* 20: 515–23.

Foster, C., Green, K., Bleda, M., Dewick, P., Evans, B., Flynn, A. & Mylan, J. 2006. *Environmental Impacts of Food Production and Consumption: A report to the Department for Environment, Food and Rural Affairs*. Manchester Business School. Defra, London.

Lehto, M., Sipilä, I., Alakukku, L. & Kymäläinen, H.-R. 2014. Water consumption and wastewaters in fresh-cut vegetable production. *Agricultural and Food Science* 23: 246–56.

Martín-Belloso, O. & Soliva-Fortuny, R. (eds). 2010. *Advances in Fresh-Cut Fruits and Vegetables Processing*. Food Preservation Technology Series. CRC Press, Florida.

Moore, L.V. & Thompson, F.E. 2015. Adults meeting fruit and vegetable intake recommendations – United States, 2013. *Morbidity and Mortality Weekly Report* 64: 709–13.

Raffo, A., Leonardi, C., Fogliano, V., Ambrosino, P., Salucci, M., Gennaro, L., Bugianesi, R., Giuffrida, F. & Quaglia, G. 2002. Nutritional value of cherry tomatoes (*Lycopersicon esculentum* Cv. Naomi F1) harvested at different ripening stages. *Journal of Agricultural and Food Chemistry* 50: 6550–6.

Roberts, C. 2014. Chapter 7: Fruit and vegetable consumption. *The Health Survey for England 2013: Volume 1*. The Health and Social Care Information Centre.

Rojas-Graü, M.J., Soliva-Fortuny, R. & Martín-Belloso, O. 2009. Edible coatings to incorporate active ingredients to fresh-cut fruits: a review. *Food Science & Technology* 20: 438–47.

Stuart, D. 2011. 'Nature' is not guilty: foodborne illness and the industrial bagged salad. *Sociologia Ruralis* 51: 158–74.

Van Herpen, E., Immink, V. & van den Puttelaar, J. 2016. Organics unpacked: the influence of packaging on the choice for organic fruits and vegetables. *Food Quality and Preference* 53: 90–6.

Williams, H. & Wikström, F. 2011. Environmental impact of packaging and food losses in a life cycle perspective: a comparative analysis of five food items. *Journal of Cleaner Production* 19: 43–8.

World Carrot Museum www.carrotmuseum.co.uk

Chapter 5

Braun, D.R., Harris, J.W.K., Levin, N.E., McCoy, J.T., Herries, A.I.R., Bamford, M.K., Bishop, L.C., Richmond, B.G., Kibunjia, M. & Klein, R.G. 2010. Early hominin diet included diverse terrestrial and aquatic animals 1.95 mya in East Turkana, Kenya. *PNAS* 107: 10002–7.

Bruinsma, J. (ed.). 2003. *World Agriculture: Towards 2015/2030 – An FAO Perspective*. Earthscan Publications Ltd.: London.

Carmody, R.N. & Wrangham, R.W. 2009. The energetic significance of cooking. *Journal of Human Evolution* 57: 379–91.

Carmody, R.N., Weintraub, G.S. & Wrangham, R.W. 2011. Energetic consequences of thermal and nonthermal food processing. *PNAS* 108: 19199–203.

Dunn, R. 2012. Human ancestors were nearly all vegetarians. *Scientific American*.https://blogs.scientificamerican.com/guest-blog/human-ancestors-were-nearly-all-vegetarians

EFSA Panel on Biological Hazards (BIOHAZ). 2013. Scientific Opinion on the public health risks related to mechanically separated meat (MSM) derived from poultry and swine. *EFSA Journal* 11: 3137.

Evershed, R.P., Copley, M.S., Dickson, L. & Hansel, F.A. 2008. Experimental evidence for the processing of marine animal products and other commodities containing polyunsaturated fatty acids in pottery vessels. *Archaeometry* 50: 101–13.

Hui, Y.H. 2012. *Handbook of Meat and Meat Processing, Second Edition*. Taylor & Francis Group, CRC Press: Boca Raton.

Mendonça, R.C.S., Gouvêa, D.M., Hungaro, H.M., Sodré, A. de F. & Querol-Simon, A. 2013. Dynamics of the yeast flora in artisanal country style and industrial dry cured sausage (yeast in fermented sausage). *Food Control* 29: 143–8.

Milton, K. 1999. A hypothesis to explain the role of meat-eating in human evolution. *Evolutionary Anthropology Issues News and Reviews*: 11–21.

Newman, P.B. 1981. The separation of meat from bone – a review of the mechanics and the problems. *Meat Science* 5: 171–200.

Nychas, G.J.E. & Arkoudelos, J.S. 1990. Staphylococci: their role in fermented sausages. *Journal of Applied Bacteriology* 167S–188S.

Tuomisto, H.L. & Teixeira de Mattos, M.J. 2011. Environmental impacts of cultured meat production. *Environmental Science & Technology*, 45: 6117–23.

Chapter 6

Cordain, L., Eaton, S.B., Sebastian, A., Mann, N., Lindeberg, S., Watkins, B.A., O'Keefe, J.H. & Brand-Miller, J. 2005. Origins and evolution of the Western diet: health implications for the 21st century. *The American Journal of Clinical Nutrition* 81: 341–54.

Hätönen, K.A., Virtamo, J., Eriksson, J.G., Sinkko, H.K., Sundvall, J.E. & Valsta, L.M. 2011. Protein and fat modify the glycaemic and insulinaemic responses to a mashed potato-based meal. *The British Journal of Nutrition* 106: 248–53.

Inglett, G.E. 1976. A history of sweeteners – natural and synthetic. *Journal of Toxicology and Environmental Health* 2: 207–14.

Nielsen Global Survey of Snacking. 2014. *Global snack food sales reach $374 billion annually.* www.nielsen.com

Paz-Filho, G., Mastronardi, C., Wong, M.-L. & Licinio, J. 2012. Leptin therapy, insulin sensitivity, and glucose homeostasis. *Indian Journal of Endocrinology and Metabolism* 16: S549–55.

Reed, D.R. & Xia, M.B. 2015. Recent advances in fatty acid perception and genetics. *Advances in Nutrition* 6: 353S–360S.

Rizkalla, S.W. 2010. Health implications of fructose consumption: A review of recent data. *Nutrition & Metabolism* 7: PMC2991323.

Scarborough, P., Bhatnagar, P., Wickramasinghe, K.K., Allender, S., Foster, C. & Rayner, M. 2011. The economic burden of ill health due to diet, physical inactivity, smoking, alcohol and obesity in the UK: an update to 2006–2007 NHS costs. *Journal of Public Health (Oxford, England)* 33: 527–35.

Smith, C.W. 1995. *Crop Production: Evolution, History and Technology.* John Wiley & Sons Inc, New York.

Struck, S., Jaros, D., Brennan, C.S. & Rohm, H. 2014. Sugar replacement in sweetened bakery goods. *International Journal of Food Science and Technology* 49: 1963–76.

Chapter 7

Aguiar, M. & Hurst, E. 2006. *Measuring Trends in Leisure: The Allocation of Time over Five Decades.* Federal Reserve Bank of Boston.

Akerwi, A., Crichton, G.E. & Hébert, J.R. 2015. Consumption of ready-made meals and increased risk of obesity: findings from the Observation of Cardiovascular Risk Factors in Luxembourg (ORISCAV-LUX) study. *British Journal of Nutrition* 113: 270–7.

Cardello, H. & Garr, D. 2009. *Stuffed.* HarperCollins Publishers, New York.

Goldin, C. 1977. Female labor force participation: the origin of black and white differences, 1870 and 1880. *Journal of Economic History* 37: 87–108.

Henry, C.J.K. & Heppell, N. 2002. Nutritional losses and gains during processing: future problems and issues. *Proceedings of the Nutrition Society* 61: 145–8.

Howard, S., Adams, J. & White, M. 2012. Nutritional content of supermarket ready meals and recipes by television chefs in the United Kingdom: cross sectional study. *BMJ* 345: e7607.

Jabs, J. & Devine, C.M. 2006. Time scarcity and food choices: An overview. *Appetite* 47: 196–204.

Parker, K. & Wang, W. 2013. *Modern Parenthood: Roles of Moms and Dads Converge as they Balance Work and Family*. Pew Research Center, Washington, D.C.

Schmidt Rivera, X. C., Espinoza Orias, N. & Azapagic, A. 2014. Life cycle environmental impacts of convenience food: Comparison of ready and home-made meals. *Journal of Cleaner Production* 73: 294–309.

Chapter 8

Athinarayanan, J., Periasamy, V.S., Alsaif, M.A., Al-Warthan, A.A. & Alshatwi, A.A. 2014. Presence of nanosilica (E551) in commercial food products: TNF-mediated oxidative stress and altered cell cycle progression in human lung fibroblast cells. *Cell Biology and Toxicology* 30: 89–100.

Dekkers, S., Krystek, P., Peters, R.J.B., Lankveld, D.P.K., Bokkers, B.G.H., van Hoeven-Arentzen, P.H., Bouwmeester, H. & Oomen, A.G. 2011. Presence and risks of nanosilica in food products. *Nanotoxicology* 5: 393–405.

Dev, B.C., Sood, S.M., DeWind, S. & Slattery, C.W. 1994. Kappa-casein and beta-caseins in human milk micelles: structural studies. *Archives of Biochemistry and Biophysics* 314: 329–36.

Hristov, P., Mitkov, I., Sirkova, D., Mehandgiiski, I. & Radoslavov, G. 2016. Measurement of casein micelle size in raw dairy cattle milk by dynamic light scattering. In *Milk Proteins – From Structure to Biological Properties and Health Aspects* (Gigli, I., ed.). InTech: 10.5772/62779.

Pérez-Esteve, É., Bernardos, A., Martínez-Mañez, R. & Barat, J.M. 2011. Recent patents in food nanotechnology. *Recent Patents on Food, Nutrition & Agriculture* 3: 172–8.

Peters, R., Kramer, E., Oomen, G.G., Herrera Rivera, Z.E., Oegema, G., Tromp, P.C., Fokkink, R., Rietveld, A., Marvin, H.J.P., Weigel, S., Peijnenburg, A.A.C.M. & Bouwmeester, H. 2012. Presence of nano-sized silica during *in vitro* digestion of foods containing silica as a food additive. *ACS Nano* 6: 2441–51.

Rai, M., Ribeiro, C., Mattoso, L. & Duran, N. (eds). 2015. *Nanotechnologies in Food and Agriculture*. Springer International Publishing, Switzerland.

Ramos, O.L., Pereira, R.N., Martins, A., Rodrigues, R., Fuciños, C., Teixeira, J.A., Pastrana, L., Malcata, F.X. & Vicente, A.A. 2015. Design of whey protein nanostructures for incorporation and release of nutraceutical compounds in food. *Critical Reviews in Food Science and Nutrition* 57: 1377–93.

Rogers, M.A. 2016. Naturally occurring nanoparticles in food. *Current Opinion in Food Science* 7: 14–19.

Szakal, C., Roberts, S.M., Westerhoff, P., Bartholomaeus, A., Buck, N., Illuminato, I., Canady, R. & Rogers, M. 2014. Foods: Integrative consideration of challenges and future prospects. *ACS Nano* 8: 3128–35.

Yada, R.Y., Buck, N., Canady, R., DeMerlis, C., Duncan, T., Janer, G., Juneja, L., Lin, M., McClements, J., Noonan, G., Oxley, J., Sabliov, C., Tsytsikova, L., Vázquez-Campos, S., Yourick, J., Zhong, Q. & Thurmond, S. 2014. Engineered nanoscale food ingredients: evaluation of current knowledge on material characteristics relevant to uptake from the gastrointestinal tract. *Comprehensive Reviews in Food Science and Food Safety* 13: 730–44.

Chapter 9

Burkle, L.A., Marlin, J.C. & Knight, T.M. 2013. Plant-pollinator interactions over 120 years: loss of species, co-occurrence, and function. *Science* 339: 1611–15.

Garibaldi, L.A., Steffan-Dewenter, I., Winfree, R., et al. 2013. Wild pollinators enhance fruit set of crops regardless of honey bee abundance. *Science* 339: 1608–11.

Institute for The Future. 2011. *Four Futures of Food: Global Food Outlook Alternative Scenarios Briefing*. www.iftf.org/uploads/media/IFTF_SR1388_GFOFuturesFood.pdf

Manuell, R. 2017. New Food presents... 'The 10 Top Trends of 2017'. www.newfoodmagazine.com/news/28790/10-top-trends-2017/

Acknowledgements

I must first thank my publisher, Jim Martin at Bloomsbury, for giving me another opportunity to ramble on about the science of food. He is extremely supportive and straight-up in his approach, which I love. I would also like to thank Anna MacDiarmid, editor at Bloomsbury, who came up with the title for the book early in the process and it just took me a very long time to realise her brilliance! It is also thanks to Anna's efficiency and diligence that a Word document is transformed into a book – from my perspective she makes the whole process seem smooth, but I'm sure her perspective is very different. Thank you to James Watson for designing a fabulous eye-catching cover.

In terms of the development of the text, I would like to thank Sandra Banner, Alida Robey and Ian Glen for their comments on various portions of the text, which helped to improve it considerably. However, it is my sister, Jennifer Gruno, who questioned and corrected the majority of these pages, for which I am eternally grateful. With her constructive comments, she allowed me to view my text through the eyes of my target audience, a rare gift for an author. I owe enormous thanks to Catherine Best, my copy editor, who has once again helped to shape my words, clarify inconsistencies and generally make this a better piece of work.

There are many scientists who gave up their time to discuss various aspects of food science with me. The most constructive way I can thank you all, I believe, is to have represented your science accurately. I would particularly like to thank John Floros, Dean of the College of Agriculture, not only for speaking with me, but also for writing some rather pivotal papers that reshaped my thinking on food processing. I would like to thank the International Iberian Nanotechnology Laboratory for inviting me to Portugal to learn about nanotechnology and food, and in particular their Head of Life Science, Lorenzo Pastrana.

And then there are all of the fabulous people who surround me and support me by getting me out of my home office, taking my son for play dates, telling me to get off Facebook when I should be writing, giving me encouragement when I feel down and plying me with coffee or wine as necessary: thank you. To the family members we have lost since I last wrote an acknowledgement – Grandma Gwen, Grandma Terry and Jim – and to the family members still here who envelop me in love daily, thank you. The world would be a better place if everyone felt so supported to pursue their dreams.

Index